Numerical
Analysis

Books in this Series

Numerical Analysis

Charles Dixon
B.Sc., Ph.D., F.R.Met.S.

University of Dundee

Blackie Glasgow and London
Chambers Edinburgh

Blackie & Son Limited   Bishopbriggs, Glasgow G64 2NZ
5 Fitzhardinge Street, London W1H 0DL

W. & R. Chambers Limited   11 Thistle Street, Edinburgh EH2 1DG

First published   1974

Blackie (educational)   0 216 89681 9
(net)   0 216 89682 7
Chambers (educational)   0 550 75895 X
(net)   0 550 75896 8

Filmset and printed by Thomson Litho, East Kilbride, Scotland.

Preface

The importance of numerical analysis has increased considerably in the last few years. Today no university course in Mathematics is complete without at least an introduction to the subject being given. It is a subject which can be made appealing to students at various levels of sophistication and a carefully prepared introduction can be a suitable course for sixth year pupils in schools. For these pupils the subject must not be reduced to the mere mechanical evaluation of formulae. This approach makes the subject artificially dull and may do lasting harm to the pupils. It may put them off completely from studying further a subject which has so much to offer which is intellectually stimulating. Pupils must understand the theoretical aspects of the subject. However, the theory must not be over-emphasized at the expense of losing sight of the practicality of the subject. A middle course must be steered between theory and computation. It is hoped that this text will help in this aim.

Another danger, when introducing the subject, is to give the impression that everything is cut and dried so that no interesting problems remain. This too tends to kill interest in the subject. It is hoped that, in particular, the brief mention of ill-conditioning at various points will help to dispel this impression.

It must be emphasized too that numerical analysis is very closely linked up with computers. In a real situation the computer is used to carry out the (otherwise tedious) numerical computations. The reader should therefore attempt to make full use of any computing facilities available to him when doing many of the exercises given in the text.

The text has been written with the content of the relevant syllabus for sixth year studies in Scottish schools firmly in mind. It is based in part on a series of lectures given annually for several years at a school teachers' summer school run jointly by Dundee College of Education and Dundee University. It is hoped that the text may also be helpful to students attending first year courses at Colleges and Universities. Some background material which may not be familiar to all readers is included in appendices.

My thanks are due to Dr D P Thomas, University of Dundee, for many helpful discussions and to Mr A McMeeken, College of Education, Dr J M

Rushforth, University of Dundee and Mr N Smart, Aberdeen College of Education who read the initial manuscript very thoroughly and made many valuable suggestions. As a result of these discussions and the amendments made the final text should be more acceptable to the students for whom it is intended.

Contents

Introduction

‖

IN AN ATTEMPT TO GAIN A CLEARER UNDERSTANDING OF A PROBLEM IN
the physical or biological sciences, engineering or the social sciences, the
applied mathematician first tries to contruct a mathematical model of
the given situation. He tries to express the essential features of the problem
in terms of a series of equations or inequations involving the relevant
variables. These may be temperature and pressure in a physical problem,
or buying and selling prices in an economics problem. Once he has
obtained this mathematical model (and this is normally a very difficult
task) his next step is to solve the equations or inequations of the model.
If he is lucky, by using analytical methods he can perhaps obtain his
solution in terms of a formula or set of formulae. Subsequent examination
of these formulae might reveal a considerable amount of qualitative
results. For example, it might be revealed that the effects of, say, temperature are very small, while the pressure effects are relatively large. However,
quantitative results may also be required, that is, numerical values may be
required for some of the variables in the problem in certain situations.
These can be obtained by substituting specific values in the formulae
representing the solution. This process is not always as easy as it may
appear at the moment, and indeed the accuracy of our final results may
depend to a considerable extent on the way in which this is done. Choosing
the best way in a given case (and carrying it out) is a problem in numerical
analysis.

Example.   Determine the root of the equation $2x^2 + 3x - 1 = 0$ which lies
between 0 and 1.

(i) Proceeding by an analytic method we can complete the square in x on
the left-hand side of the given equation to obtain

$$2(x + \tfrac{3}{4})^2 - 2(\tfrac{3}{4})^2 - 1 = 0$$

Hence
$$(x + \tfrac{3}{4})^2 = \tfrac{17}{16}$$

Then
$$x = -\tfrac{3}{4} \pm \tfrac{\sqrt{17}}{4}$$

Therefore the required root is given by $-\frac{3}{4}+\frac{\sqrt{17}}{4}$ and this can now be evaluated to any required accuracy provided $\sqrt{17}$ can be obtained to the appropriate accuracy.

(ii) Proceeding by a numerical method we can rewrite the given equation in the form

$$x = \tfrac{1}{3}(1-2x^2)$$

Now, starting with the approximation 0 to the required root, we can obtain a better approximation to the root by putting $x = 0$ in the right-hand side of the above equation. Thus we obtain the better approximation $\tfrac{1}{3}(1-2\times 0) = \tfrac{1}{3}$. Repeating the process we obtain the still better approximation $\tfrac{1}{3}(1-2(\tfrac{1}{3})^2) = \tfrac{7}{27}$. By repeating the process often enough we can, in this case, evaluate the root to any required accuracy.

Numerical processes of this type are called *iterative processes* and will be discussed more fully in Chapter 7.

At this stage the reader may well be of the opinion that in this example the analytical method is much superior to the numerical method. However, it must be pointed out that for equations of higher degree than the third, no analytical method corresponding to the above exists, whereas iterative methods similar to the above will still be applicable for determining a root.

In some cases it may be extremely difficult (or even impossible) to obtain a solution for a given problem in terms of a formula or set of formulae using analytical methods, and in such cases it is necessary to use a numerical technique right from the start. Doing this we go straight to the quantitative results, that is, to the numerical values of our solution in various circumstances. If we compute sufficient of these, it may then also be possible to deduce some qualitative results.

But numerical analysis is not merely a tool which is to be used when all else has failed. It may be that in certain problems it is preferable to use numerical analysis from the start and to go directly to the quantitative results, even when it is known that analytical methods would lead to a solution in the form of a formula or set of formulae. This decision might be reached simply because it is easier or faster (as a computer program may already exist) to do so. Alternatively, it may be cheaper to do the complete problem on the computer using numerical techniques than first to obtain a solution using analytical methods and then to use the computer to evaluate the resulting formula or formulae.

For example, after considerable effort, it can be shown that

$$\int_a^b \frac{1}{1+x^4}\, dx = \frac{1}{4\sqrt{2}}\left[\ln\left(\frac{x^2+x\sqrt{2}+1}{x^2-x\sqrt{2}+1}\right)+2\tan^{-1}\left(\frac{x\sqrt{2}}{1-x^2}\right)\right]_a^b$$

and clearly a considerable amount of work has still to be done before the numerical value of the integral is obtained. In this case it would be preferable to evaluate the integral by numerical methods right from the start.

An advantage of numerical analysis is that in a sense the methods used are more general than are analytical methods, in that when a solution exists it can normally be computed. We can have two similar-looking problems, one of which can be solved easily analytically and the other of which requires a completely different method, or indeed may not be solvable at all analytically. A disadvantage is that using numerical techniques only one particular problem is solved at a time; and if another similar problem, with perhaps different data, is then to be solved, the whole process of solution has generally to be repeated from the beginning. However, with computers at their present stage of development this is no longer a serious problem.

It is not always true that the scientist, the engineer and the social scientist are solely interested in numerical values for their solution function. It may be that they are more interested in the effects of changes in the values of several variables on the final result. This more basic understanding can also be obtained by using numerical analysis and computers, and carrying out a series of numerical experiments. The problem is solved several times on the computer with different values for the starting data.

For example, the effect on the climate of certain areas of the earth of removing part of the polar ice-cap might be investigated by setting up a mathematical model of the general circulation of the atmosphere and then solving the resulting equations numerically with different boundary conditions corresponding to different positions of the edge of the ice-cap. Physical modelling of this situation in the laboratory is very difficult, and clearly any experimentation on the real system would be very costly and, more important, might lead to climatic disasters for certain areas.

No doubt some of the points raised above will only become clear to the reader after he has obtained some experience in numerical analysis from the subsequent sections.

Errors

2

OBTAINING NUMERICAL VALUES FOR THE SOLUTION OF A MATHEMATICAL model is only part of the problem. We shall also normally require to know just how accurate are the numerical results. This is a very important part and is often more difficult than working out the results themselves.

In calculations there are four types of error which can arise affecting the accuracy of the results. These are

1. round-off errors
2. errors due to imprecision of the given data
3. mistakes
4. errors due to the particular method used.

We shall now look at each of these in turn.

2.1. Round-off errors

When carrying out numerical calculations there is a physical limit to the number of digits we can retain in a number; that is, we must round off our numbers to a finite number of digits. The error involved in doing this is called the *round-off error*.

For example (a) $\frac{4}{3} = 1 \cdot 3333\ldots$

To four significant figures $\frac{4}{3} = 1 \cdot 333$

Round-off error $= 1 \cdot 3333\ldots - 1 \cdot 333$

$$= 0 \cdot 000333\ldots$$

(b) $\frac{5}{3} = 1 \cdot 6666\ldots$

To four significant figures $\frac{5}{3} = 1 \cdot 667$

Round-off error $= 1 \cdot 6666\ldots - 1 \cdot 667$

$$= -0 \cdot 000333\ldots$$

(c) The following table shows the results of rounding exact numbers to N significant figures.

Number	N	Rounded number	Round-off error
23·764462	5	23·764	0·000462
0·0092746	3	0·00927	0·0000046
75684·3195	3	75700	−15·6805
1·650045	5	1·6500	0·000045
0·57386	3	0·574	−0·00014
0·0003786	3	0·000379	−0·0000004
1·2545	4	1·254	0·0005
1·2545	2	1·3	−0·0455
1·25451	4	1·255	−0·00049
1·25451	3	1·25	0·00451
1·25451	2	1·3	−0·04549
1·25	2	1·2	0·05
1·35	2		

Careful study of this table will reveal the danger of rounding to fewer significant figures numbers which have already been rounded.

Now the working of a computer is such that it is convenient for the computer to carry out its calculations rounding off to a fixed number of significant figures at each stage, and indeed this is often the best approach when carrying out a calculation manually also. However, there are occasions when it may be necessary for us to use numbers which have been obtained to a fixed number of decimal places. It should be emphasized that rounding a number to a given number of significant figures will *not*, in general, give the same result as rounding this number to the same number of decimal places.

For example, the following table shows the result of rounding exact numbers to N decimal places.

Number	N	Rounded number	Round-off error
23·764462	5	23·76446	0·000002
0·0092746	3	0·009	0·0002746
75684·3195	3	75684·320	−0·0005
1·650045	5	1·65004	0·000005
0·57386	3	0·574	−0·00014
0·0003786	3	0·000	0·0003786

The results given here should be compared with the corresponding results in the previous table. In particular it should be observed that after rounding the last number in this table we do not obtain any significant figures at all.

Note. When rounding to k decimal places the round-off error must be within the limits

$$\pm \tfrac{1}{2}10^{-k}$$

For example, when rounding to 3 decimal places the round-off error must lie between 0·0005 and −0·0005, that is between $\tfrac{1}{2}10^{-3}$ and $-\tfrac{1}{2}10^{-3}$.

2.2. Imprecision of the given data

If the data are obtained experimentally, then they are of course only known within the limits of experimental error (which can normally be estimated), and this will limit the accuracy of the results of any subsequent calculations. This apparently obvious fact—that the accuracy of results is limited by the accuracy of any initial data—is one that is frequently ignored by beginners.

2.3. Mistakes

Mistakes occur by misuse of the methods being applied. They are errors which are created by the person performing the calculations. A common mistake is to invert the order of two digits occurring in a number. For example, it is very easy to use the number 62381 instead of the number 63281. When doing calculations, as many checks as possible should be incorporated in the method itself so that any mistakes come quickly to light. Mere repetition of a calculation is itself not a good check as it is quite likely that the same mistake will be repeated.

If the computations are to be carried out on a computer, then it is not necessary to incorporate checks in the method, provided the computer program has been fully checked and it is known that the method being used is suitable for the particular problem being tackled.

2.4. Errors of the method

Errors of the method are due to replacing an exact process by an approximate one. For example, we introduce an error if we use only a finite number of terms from an infinite series expansion. This error is called a *truncation error*, that is, the error due to truncating the series somewhere.

For example $\sin x$ can be expressed as the infinite series expansion*

$$x-\frac{x^3}{3!}+\frac{x^5}{5!}-\frac{x^7}{7!}+\frac{x^9}{9!}-\frac{x^{11}}{11!}+\dots$$

---

* For positive integers n, $n! = 1.2.3.\dots(n-1).n$.

and when x is small the sum of the first three terms, namely $x - \dfrac{x^3}{3!} + \dfrac{x^5}{5!}$,

will give a good approximation to $\sin x$. The truncation error is then the sum of the remaining terms of the infinite series expansion namely

$$-\frac{x^7}{7!} + \frac{x^9}{9!} - \frac{x^{11}}{11!} + \cdots\cdots$$

EXERCISES

1. Round-off the following (exact) numbers to the number of significant figures indicated and state the round-off error introduced in each case.
   (i) 1·7514 to 3 significant figures
   (ii) 0·04795 to 3 significant figures
   (iii) 349·1862 to 3 significant figures
   (iv) 29·34625 to 4 significant figures
   (v) 0·486251 to 4 significant figures
   (vi) 1·6485003 to 4 significant figures
   (vii) 0·0004985 to 3 significant figures
   (viii) 5·0032 to 2 significant figures

2. Round-off the numbers given in exercise 1 to the number of decimal places indicated and state the round-off error introduced in each case.
   (i) 3 decimal places
   (ii) 3 decimal places
   (iii) 3 decimal places
   (iv) 4 decimal places
   (v) 4 decimal places
   (vi) 4 decimal places
   (vii) 3 decimal places
   (viii) 2 decimal places

2.5. Absolute error, relative error, precision and accuracy

2.5.1. Absolute error and relative error

If x represents an approximation to the value X and e is the error in this approximation (due to whatever source), then

$$X = x + e$$

The numerical value of the error e, denoted by $|e|$, is called the *absolute error* and $\left|\dfrac{e}{X}\right|$ is called the *relative error*.

For example, 27 may be used as an approximation to the (exact) number 26·76. In this case $X = 26\cdot76$ and $x = 27$. The error e (due to round-off in this case) is $-0\cdot24$, the absolute error is $0\cdot24$ and the relative error is $\frac{0\cdot24}{26\cdot76} = 0\cdot009$ correct to three decimal places or one significant figure.

Each of the concepts of absolute error and relative error is useful in different circumstances. For example, to say that the absolute error in some number is 0·0004 does not mean very much unless we know just how big the number is itself. If the number is approximately 2, the above error may not be too important. The relative error is then only 0·0002, that is 2 parts in 10 000. However, if the given number is approximately 0·002, the given absolute error is almost certainly more important. The relative error is 2 parts in 10. But if the number being estimated is itself zero, any error at all (however small) would give an infinite relative error. In such a case the absolute error would be more meaningful.

In general we shall see that, when adding and subtracting, the absolute errors are more useful, whereas when multiplying and dividing the relative errors are more useful.

2.5.2. *Accuracy and precision*

Hitherto accuracy and precision have been used almost as synonyms. We shall now draw a distinction between the meanings of these two words.

The *precision* of a measurement depends upon the least unit of measurement employed. Thus measurements of time correct to the nearest second are equally precise, even although they may not involve the same number of significant figures. The precision of a measurement is related to the absolute error in its value. The smaller the absolute error the greater the precision.

For example (*a*) 6517·2 seconds and 3·4 seconds each measured correct to the nearest 0·1 second are equally precise. (*b*) 1072 seconds and 5 seconds each measured to the nearest second are also equally precise. Both have a possible absolute error 0·5 second. (*c*) 1072 seconds measured to the nearest second is less precise than 5·1 seconds measured to the nearest 0·1 second. The least unit of measurement used in the measurement 5·1 seconds (0·1 second) and the possible absolute error in this value (0·05 second) are both less than the corresponding quantities for the measurement 1072 seconds (namely 1 second and 0·5 second).

The *accuracy* of a measurement depends on the number of significant figures in its value (and not on the least unit of measurement). Thus measurements of time with the same number of significant figures will be equally accurate, even although the least units of measurement used may be very different, for example 1 second, 1 hour, 1 day. The accuracy of a measurement is related to the relative error in its value. The smaller the relative error the greater the accuracy.

For example (*a*) 23·7 seconds (to the nearest 0·1 second), 2·37 hours (to the nearest 0·01 hour) and 237 days (to the nearest day) are all equally accurate measurements (although their precisions are very different). Note that the relative error in each of the above measurements is the same, namely $\frac{5}{2370}$.

(b) The measurement 132·5 days (to the nearest 0·1 day) is more accurate than the measurement 2·5 seconds (to the nearest 0·1 second). Note, however, that 132·5 days is considerably less precise than 2·5 seconds.

2.6. The effects of error on the basic operations of arithmetic

In what follows it will be assumed that the relative errors occurring in the numbers used in the basic operations are small. If this assumption is not valid, that is, if the error in a number is of the same order of magnitude as the number itself, then it is pointless to continue with any error analysis such as the following.

2.6.1. Addition (and subtraction)

Example. Two lengths measured correct to the nearest 0·1 mm are 3·2 mm and 1·6 mm. What is the best estimate we can obtain, from these two measurements, of the sum of the two (exact) lengths? Consider the error in the estimate.

The first length lies somewhere in the interval (3·2 ± 0·05) mm, that is between 3·15 mm and 3·25 mm.

The second length lies somewhere in the interval (1·6 ± 0·05) mm, that is between 1·55 mm and 1·65 mm.

Hence the minimum value for the sum is (3·15 + 1·55) mm = 4·70 mm and the maximum value for the sum is (3·25 + 1·65) mm = 4·90 mm. Thus the value 4·8 mm, obtained by simply adding 3·2 mm and 1·6 mm, looked upon as an approximation to the sum of the two exact lengths has a possible absolute error of 0·1 mm. Note that this maximum absolute error of 0·1 mm is equal to the sum of the maximum absolute errors in the original measurements (each of which was 0·05 mm). Thus the maximum absolute error in the sum of two measurements is equal to the sum of the maximum absolute errors in the two measurements. The estimate of the sum can be given as 5 mm correct to the nearest millimetre or as (4·8 ± 0·1) mm. The result should not be given simply as 4·8 mm, as this gives the impression that it is correct to one decimal place.

In general, if x_1 and x_2 are approximations to the true values X_1 and X_2 respectively, and e_1 and e_2 are the corresponding errors in these approximations, then

$$X_1 = x_1 + e_1 \quad \text{and} \quad X_2 = x_2 + e_2.$$

Therefore $X_1 + X_2 = (x_1 + x_2) + (e_1 + e_2).$

Thus the sum $(x_1 + x_2)$ of the two approximate values approximates the sum $(X_1 + X_2)$ of the two exact values with an error $(e_1 + e_2)$, the sum of the individual errors.

Then
$$|(X_1 + X_2) - (x_1 + x_2)| = |e_1 + e_2|$$
$$\leqslant |e_1| + |e_2|*$$

Thus the absolute error in the sum of two numbers is less than or equal to the sum of the absolute errors in the individual numbers.

Now if the error e_1 is due to rounding-off X_1 to k_1 decimal places and the error e_2 is due to rounding-off X_2 to k_2 decimal places

$$-\tfrac{1}{2}10^{-k_1} \leqslant e_1 \leqslant \tfrac{1}{2}10^{-k_1} \quad \text{and} \quad -\tfrac{1}{2}10^{-k_2} \leqslant e_2 \leqslant \tfrac{1}{2}10^{-k_2}.$$

That is
$$|e_1| \leqslant \tfrac{1}{2}10^{-k_1} \quad \text{and} \quad |e_2| \leqslant \tfrac{1}{2}10^{-k_2}.$$

Hence
$$|e_1 + e_2| \leqslant \tfrac{1}{2}10^{-k_1} + \tfrac{1}{2}10^{-k_2}.$$

Hence the error $(e_1 + e_2)$ in the approximation $(x_1 + x_2)$ to the true value $(X_1 + X_2)$ lies within the limits $\pm(\tfrac{1}{2}10^{-k_1} + \tfrac{1}{2}10^{-k_2})$.

In particular, as will commonly happen in practice, if both the given numbers are rounded to the same number k (say) decimal places, then $k_1 = k_2 = k$ and the error in $(x_1 + x_2)$ lies within the limits $\pm 10^{-k}$. Thus the approximation may be as much as one unit out in the kth decimal place.

Example. The numbers a and b when rounded-off to 3 decimal places are 3·724 and 4·037 respectively. Evaluate an approximation to $(a + b)$ and consider the error involved.

An approximation to $(a + b)$ is given by $(3·724 + 4·037) = 7·761$. The error due to round-off lies between -10^{-3} and 10^{-3}, that is between $-0·001$ and $0·001$. Hence the true value of $(a + b)$ lies between 7·760 and 7·762. To two decimal places this is 7·76 and, while the third decimal place is not completely determined, we at least know that it is 0, 1 or 2. The result should not be quoted as 7·761 as this gives the impression that it is correct to 3 decimal places. It should be quoted as 7·76 to 2 decimal places or as $7·761 \pm 0·001$.

Example. The numbers a and b, when rounded to four significant digits, are 23·86 and 0·01762 respectively. Evaluate an approximation to $(a + b)$ and discuss the error involved.

An approximation to $(a + b)$ is given by $(23·86 + 0·01762) = 23·87762$. The round-off error lies between $\pm(\tfrac{1}{2}10^{-2} + \tfrac{1}{2}10^{-5})$, that is between 0·005005 and $-0·005005$.

---

* If a and b are any two real numbers, positive or negative, then $|a + b| \leqslant |a| + |b|$.

Hence the true value of $(a+b)$ lies between 23·882625 and 23·872615. Thus to four significant digits the true result is either 23·88 or 23·87. To three significant digits the result is 23·9. The result can also be given as $23·878 \pm 0·005$.

For subtraction we have

$$X_1 - X_2 = (x_1 - x_2) + (e_1 - e_2)$$

and so
$$|(X_1 - X_2) - (x_1 - x_2)| = |e_1 - e_2|$$
$$\leqslant |e_1| + |e_2|*$$

Then if e_1 and e_2 are rounding errors such that $|e_1| \leqslant \frac{1}{2}10^{-k_1}$ and $|e_2| \leqslant \frac{1}{2}10^{-k_2}$ we have

$$|e_1 - e_2| \leqslant \frac{1}{2}10^{-k_1} + \frac{1}{2}10^{-k_2}.$$

Hence the error in the approximation $(x_1 - x_2)$ to the true value $(X_1 - X_2)$ must also lie within the limits $\pm(\frac{1}{2}10^{-k_1} + \frac{1}{2}10^{-k_2})$ and **not** the limits $\pm(\frac{1}{2}10^{-k_1} - \frac{1}{2}10^{-k_2})$.

For example,

32·75 rounded to 1 decimal place (three significant figures) is 32·8.
Round-off error is 0·05.
1·125 rounded to two decimal places (three significant figures) is 1·12.
Round-off error is $-0·005$.
Subtracting these rounded numbers gives $(32·8 - 1·12) = 31·68$.
Subtracting the exact numbers gives 31·625.
Hence, regarding 31·68 as an approximation to 31·625, the error is

$$31·625 - 31·68 = -0·055$$
$$= -(0·05 + 0·005)$$
$$= -(\tfrac{1}{2}10^{-1} + \tfrac{1}{2}10^{-2}).$$

In particular, if two numbers are rounded to the same number k of decimal places, the limits for the error in their difference are $\pm 10^{-k}$.

Example. The numbers a and b when rounded to 3 decimal places are 3·724 and 2·251 respectively and the numbers c and d when rounded to 4 significant digits are 23·86 and 0·01762 respectively. Evaluate approximations to (i) $b-a$ and (ii) $c-d$ and discuss the error in each approximation.

---

* If a and b are any two real numbers, positive or negative then $|a-b| \leqslant |a| + |b|$.

(i) Approximation to $(b-a)$ is given by $(2 \cdot 251 - 3 \cdot 724) = -1 \cdot 473$.
Round-off error lies between $-0 \cdot 001$ and $+0 \cdot 001$.
Hence true value of $(b-a)$ lies between $-1 \cdot 474$ and $-1 \cdot 472$.
To two decimal places this is $-1 \cdot 47$ and the third decimal figure is 2, 3 or 4.
The result may be stated as $-1 \cdot 473 \pm 0 \cdot 001$.

(ii) Approximation to $(c-d)$ is given by $(23 \cdot 86 - 0 \cdot 01762) = 23 \cdot 84238$.
Round-off error lies between $\pm \frac{1}{2}(10^{-2} + 10^{-5}) = \pm 0 \cdot 005005$.
Hence true value of $(c-d)$ lies between $23 \cdot 847385$ and $23 \cdot 837375$.
Thus to four significant figures the true value will be either $23 \cdot 85$ or $23 \cdot 84$.
To three significant figures the result is $23 \cdot 8$.
The result could also be given as $23 \cdot 842 \pm 0 \cdot 005$.

Above, we have seen that when two numbers which have been rounded to k decimal places are added, the resulting error in the sum lies between the limits $\pm 10^{-k}$. If now we add or subtract a third number, also rounded to k decimal places, we will of course introduce a further error. The total error now will lie within the limits $\pm(10^{-k} + \frac{1}{2}10^{-k}) = \pm \frac{3}{2}10^{-k}$. Similarly, if n numbers (positive or negative) rounded to k decimal places are added, the resulting error will lie within the limits $\pm \frac{n}{2}10^{-k}$.

In particular, in the case of 10 numbers all rounded to the second decimal place say (that is each with an error within the limits $\pm 0 \cdot 005$) the resulting error will be within the limits $\pm 0 \cdot 05$, that is a whole order of magnitude greater than the individual errors.

Example. The numbers a, b, c and d when rounded to 3 decimal places are $3 \cdot 724, 2 \cdot 251, 4 \cdot 037$ and $1 \cdot 939$ respectively. Evaluate approximations to

$$\text{(i)} \quad a+b-c \quad \text{and} \quad \text{(ii)} \quad a+b-c-d$$

and discuss the error in each approximation.

(i) $(a+b-c)$ is approximated by $(3 \cdot 724 + 2 \cdot 251 - 4 \cdot 037) = 1 \cdot 938$.
Round-off error lies between $\pm \frac{3}{2}10^{-3}$, that is between $\pm 0 \cdot 0015$.
Hence true value of $(a+b-c)$ lies between $1 \cdot 9365$ and $1 \cdot 9395$.
To two decimal places this is $1 \cdot 94$.

(ii) $(a+b-c-d)$ is approximated by
$$(3 \cdot 724 + 2 \cdot 251 - 4 \cdot 037 - 1 \cdot 939) = -0 \cdot 001.$$
Round-off error lies between $\pm 2 \times 10^{-3}$, that is between $\pm 0 \cdot 002$.
Hence true value of $(a+b-c-d)$ lies between $-0 \cdot 003$ and $+0 \cdot 001$.

Thus in this case we do not even have one significant figure in our result although the numbers we started with all had four significant figures. Indeed we are not even certain as to the sign of the true result. This situation has arisen because effectively we have had to evaluate the difference of two very nearly equal numbers $(a+b)$ and $(c+d)$. This is a situation which always creates difficulties in numerical analysis. If it arises in practice, it may be that it can be avoided by employing an alternative method to solve the original problem (see §2.7). If this is not convenient, then it will be necessary to use many more significant figures in the calculations than are required in the final result, in order to allow for the inevitable loss of significant digits when the subtraction is carried out.

2.6.2. Multiplication

In the same notation as before

$$X_1 X_2 = (x_1 + e_1)(x_2 + e_2)$$
$$= x_1 x_2 + e_1 x_2 + e_2 x_1 + e_1 e_2$$
$$\simeq x_1 x_2 + e_1 x_2 + e_2 x_1$$

since the product $e_1 e_2$ can be neglected compared with the terms $e_1 x_2$ and $e_2 x_1$. Then the error in the approximation $x_1 x_2$ to $X_1 X_2$ is (approximately) $(e_1 x_2 + e_2 x_1)$.

Also, the approximate absolute error is

$$|e_1 x_2 + e_2 x_1| \leqslant |e_1 x_2| + |e_2 x_1|$$
$$= |x_2||e_1| + |x_1||e_2|.$$

When e_1 and e_2 are due to rounding to k_1 and k_2 decimal places respectively, we have

$$|e_1 x_2 + e_2 x_1| \leqslant |x_2|\tfrac{1}{2}10^{-k_1} + |x_1|\tfrac{1}{2}10^{-k_2}.$$

Further if $k_1 = k_2 = k$, that is if both numbers have been rounded to k decimal places, we have

$$|e_1 x_2 + e_2 x_1| \leqslant (|x_1| + |x_2|)\tfrac{1}{2}10^{-k}.$$

The approximate relative error in the product $x_1 x_2$ is

$$\frac{|e_1 x_2 + e_2 x_1|}{|X_1 X_2|} \simeq \frac{|e_1 x_2 + e_2 x_1|}{|x_1 x_2|}$$
$$= \left|\frac{e_1}{x_1} + \frac{e_2}{x_2}\right|$$
$$\leqslant \left|\frac{e_1}{x_1}\right| + \left|\frac{e_2}{x_2}\right|.$$

That is, the approximate relative error in the product of two numbers is less than or equal to the sum of the relative errors in the two numbers.

Example. The numbers a and b when rounded to 3 decimal places are 4·701 and 0·832 respectively. Evaluate an approximation to ab and discuss the error involved.

ab is approximated by $(4·701)(0·832) = 3·911232$ (using long multiplication).

The approximate absolute error $\leqslant (4·701 + 0·832)\frac{1}{2}10^{-3} \simeq 0·0028$.

Hence the true value of ab to 3 decimal places lies between 3·908 and 3·914. This is normally stated as 3·91 to two decimal places (or 3 significant figures) or as $3·911 \pm 0·003$.

Alternatively, since the relative error in $4·701 \leqslant \dfrac{0·0005}{4·701} \simeq 0·00011$ and the relative error in $0·832 \leqslant \dfrac{0·0005}{0·832} \simeq 0·00060$, the approximate relative error in their product $\leqslant 0·00011 + 0·00060 = 0·00071$.

Hence the approximate absolute error in the product $\leqslant 0·00071 \ (3·9)$
$$\simeq 0·0028$$
which agrees with what was obtained above.

Example. The numbers a and b when rounded to four significant digits are 37·26 and 0·02146 respectively. Evaluate an approximation to ab and discuss the error involved.

ab is approximated by $(37·26)(0·02146) = 0·7995996$ (by long multiplication).

Approximate absolute error $\leqslant \frac{1}{2}(37·26 \times 10^{-5} + 0·02146 \times 10^{-2})$
$$\simeq 0·00029.$$
Hence true value of ab lies between 0·7999 and 0·7993 when rounded to four significant figures; that is, to three significant figures the true value may be either 0·800 or 0·799; to two significant figures the value is 0·80. The result may also be stated as $0·7996 \pm 0·0003$.

Considering the product $x_1 x_2 x_3$ of three inexact numbers we obtain, in the same way as above, that

the approximate absolute error $\leqslant |x_1 x_2||e_3| + |x_2 x_3||e_1| + |x_3 x_1||e_2|$

and that the approximate relative error $\leqslant \left|\dfrac{e_1}{x_1}\right| + \left|\dfrac{e_2}{x_2}\right| + \left|\dfrac{e_3}{x_3}\right|.$

From these two results it should be clear that, when considering the product of three numbers, it is better to work in terms of relative error rather than absolute error (even when it is the absolute error which is finally required).

Example. The numbers a, b and c when rounded to 3 decimal places are 4·701, 0·832 and 2·413 respectively. Evaluate an approximation to abc and discuss the error involved.

abc is approximated by $(4·701)(0·832)(2·413) = 9·437802816$. (As we shall see it is a foolish waste of time to retain so many digits in the approximation.)

Relative error in $4·701 \leqslant \dfrac{0·0005}{4·701} \simeq 0·00011$.

Relative error in $0·832 \leqslant \dfrac{0·0005}{0·832} \simeq 0·00060$.

Relative error in $2·413 \leqslant \dfrac{0·0005}{2·413} \simeq 0·00021$.

Hence approximate relative error in product

$$\leqslant 0·00011 + 0·00060 + 0·00021$$
$$= 0·00092.$$

Hence approximate absolute error in product

$$\leqslant 0·00092 \, (9·4)$$
$$\simeq 0·0086.$$

Thus the true value of the product lies between 9·429 and 9·446 when rounded to four significant figures. Therefore the true value of the product is 9·4 correct to two significant figures. The result can also be given as $9·438 \pm 0·009$ or, better, as $9·44 \pm 0·01$.

In the same way as above, it can be shown that the approximate relative error in the product $x_1 x_2 x_3 \ldots x_n$ is less than or equal to the sum of the relative errors in $x_1, x_2, x_3, \ldots, x_n$ for any positive integer n. In particular, the approximate relative error in x_1^n, in which n is a positive integer, is n times the relative error in x_1. This last result can be generalized to give the result that the approximate relative error in x_1^n for any real n is $|n|$ times the relative error in x_1.

Example. The number 7·36 is a correctly rounded approximation to the number a. Using tables of square roots obtain as accurate an approximation as possible to $\sqrt{a}$.

The approximate relative error when $\sqrt{7\cdot36}$ is used as an approximation to $\sqrt{a}$ is $\frac{1}{2}\left(\dfrac{0\cdot005}{7\cdot36}\right) \simeq 0\cdot00034$.

From (5 figure) tables $\sqrt{7\cdot36} = 2\cdot7129$ correct to four decimal places. Hence the approximate absolute error when $\sqrt{7\cdot36}$ is used as an approximation to $\sqrt{a}$ is $(0\cdot00034 \times 2\cdot7) \simeq 0\cdot001$.

Therefore $\sqrt{a}$ lies between $\sqrt{7\cdot36}+0\cdot001$ and $\sqrt{7\cdot36}-0\cdot001$.

Hence $\sqrt{a} = 2\cdot713 \pm 0\cdot001$.

Correct to two decimal places $\sqrt{a}$ is $2\cdot71$.

(Note that, correct to three decimal places $\sqrt{7\cdot365} = 2\cdot714$ and $\sqrt{7\cdot355} = 2\cdot712$).

2.6.3. Division

Again, using the same notation, we have

$$\frac{X_1}{X_2} = \frac{x_1+e_1}{x_2+e_2} = \frac{x_1}{x_2}+\left(\frac{x_1+e_1}{x_2+e_2}-\frac{x_1}{x_2}\right)$$

$$= \frac{x_1}{x_2}+\frac{x_2(x_1+e_1)-x_1(x_2+e_2)}{x_2(x_2+e_2)}$$

$$= \frac{x_1}{x_2}+\frac{x_2e_1-x_1e_2}{x_2(x_2+e_2)}.$$

Then the error involved in using $\dfrac{x_1}{x_2}$ as an approximation to $\dfrac{X_1}{X_2}$ is

$$\frac{x_2e_1-x_1e_2}{x_2(x_2+e_2)} = \left(\frac{e_1}{x_2}-\frac{x_1}{x_2^2}e_2\right)\left(1+\frac{e_2}{x_2}\right)^{-1}$$

$$= \left(\frac{e_1}{x_2}-\frac{x_1}{x_2^2}e_2\right)\left(1-\frac{e_2}{x_2}+\dots\right)$$

on application of the binomial theorem.*

Now, since e_1 and e_2 are small, so that products of at least two of them (for example e_1e_2, $e_1e_2^2$, etc.) can be neglected, we see that this error is approximately $\left(\dfrac{e_1}{x_2}-\dfrac{x_1e_2}{x_2^2}\right)$.

Thus the approximate absolute error

$$\leqslant \left|\frac{e_1}{x_2}\right|+\left|\frac{x_1e_2}{x_2^2}\right|.$$

---

* See Appendix 1.

The relative error is

$$\left|\frac{x_2e_1 - x_1e_2}{x_2(x_2 + e_2)}\right| \bigg/ \left|\frac{X_1}{X_2}\right| = \left|\frac{X_2}{X_1}\left(\frac{x_2e_1 - x_1e_2}{x_2X_2}\right)\right|$$

$$= \left|\frac{e_1}{X_1} - \frac{x_1}{X_1}\frac{e_2}{x_2}\right| = \left|\frac{e_1}{x_1}\left(1 + \frac{e_1}{x_1}\right)^{-1} - \frac{e_2}{x_2}\left(1 + \frac{e_1}{x_1}\right)^{-1}\right|$$

$$= \left|\frac{e_1}{x_1}\left(1 - \frac{e_1}{x_1} + \cdots\right) - \frac{e_2}{x_2}\left(1 - \frac{e_1}{x_1} + \cdots\right)\right|^{*}$$

$$\simeq \left|\frac{e_1}{x_1} - \frac{e_2}{x_2}\right| \leqslant \left|\frac{e_1}{x_1}\right| + \left|\frac{e_2}{x_2}\right|.$$

Hence (as for multiplication) the approximate relative error after division is less than or equal to the sum of the individual relative errors.

From the above results for absolute error and relative error it should be clear that it is easier to use relative error when considering division.

Example. The numbers a and b, when rounded to four significant digits are 37·26 and 0·05371 respectively. Evaluate an approximation to $\frac{a}{b}$ and discuss the error involved.

An approximation to $\frac{a}{b}$ is given by $\frac{37·26}{0·05371} = 693·725\ldots$

The approximate relative error $\leqslant \dfrac{0·005}{37·26} + \dfrac{0·000005}{0·05371} \simeq 0·00023.$

Hence the approximate absolute error in the quotient

$$\leqslant 0·00023\,(690) \simeq 0·16.$$

Therefore the true value of a/b lies between 693·9 and 693·6 when rounded to four significant figures. Hence, to three significant figures a/b is 694.

Note that, as a consequence of the above results for multiplication and division, the approximate relative error in the expression $\dfrac{x_1x_2}{x_3}$ is less than or equal to the sum of the relative errors in x_1, x_2 and x_3.

Example. The numbers a, b and c when rounded to three decimal places are 4·701, 0·832 and 2·413 respectively. Evaluate approximations to (i) ab/c and (ii) $1/ab$ and discuss the error in each of these approximations.

---

*See Appendix 1.

(i) $\dfrac{ab}{c}$ is approximated by $\dfrac{(4\cdot701)(0\cdot832)}{2\cdot413} = 1\cdot62090\ldots$

$$\text{The approximate relative error} \leqslant \frac{0\cdot0005}{4\cdot701} + \frac{0\cdot0005}{0\cdot832} + \frac{0\cdot0005}{2\cdot413}$$

$$\simeq 0\cdot00011 + 0\cdot00060 + 0\cdot00021$$

$$= 0\cdot00092.$$

Hence the approximate absolute error $\leqslant 0\cdot00092\,(1\cdot6) \simeq 0\cdot0015$. Therefore the true value of ab/c lies between $1\cdot619$ and $1\cdot622$ when rounded to four significant figures.

Thus the true value is $1\cdot62$ correct to three significant figures. The result can also be given as $1\cdot621 \pm 0\cdot002$.

(ii) $\dfrac{1}{ab}$ is approximated by $\dfrac{1}{(4\cdot701)(0\cdot832)} = 0\cdot255674\ldots$

The approximate relative error $\leqslant 0\cdot00011 + 0\cdot00060 = 0\cdot00071$. (Note that this is the same as the relative error in the product ab itself, since the numerator 1 in the quotient $1/ab$ is exact and so has no error.)

Hence the approximate absolute error $\leqslant 0\cdot00071\,(0\cdot26) \simeq 0\cdot00018$. Therefore the true value of $1/ab$ lies between $0\cdot2555$ and $0\cdot2559$ when rounded to four decimal places.

Hence the true value of $1/ab$ may be given as $0\cdot256$ correct to three decimal places or as $0\cdot2557 \pm 0\cdot0002$.

2.6.4. Error in functional evaluation

If e is the error in the approximation x to the number X so that $X = x + e$ then, if e_f denotes the error when a function f is evaluated (exactly) at x instead of at X, we have

$$f(X) = f(x) + e_f.$$

Therefore $\quad e_f = f(X) - f(x)$

$$= f(x+e) - f(x)$$

$$= [f(x) + ef'(x) + \tfrac{1}{2}e^2 f''(x) + \cdots] - f(x)$$

on expanding $f(x+e)$ in a Taylor series.*

Therefore $\qquad e_f = ef'(x) + \tfrac{1}{2}e^2 f''(x) + \cdots$

Hence if e is small (and the second and higher derivatives of f evaluated at x are not excessively large) we see that $e_f \simeq ef'(x)$.

---

* See Appendix 1.

Thus $$|e_f| \simeq |e||f'(x)|.$$

Because only the first one or two significant figures in an error value are useful, it will normally only be necessary to evaluate $f'(x)$ correct to two significant figures.

Example. The example on page 15 obtaining an approximation to $\sqrt{a}$ is an example of error in function evaluation.

Here $$f(x) \equiv \sqrt{x}$$

and so $$|e_f| \simeq |e||f'(x)| = |e||\tfrac{1}{2}x^{-\frac{1}{2}}|.$$

Hence the approximate absolute error in $\sqrt{7.36} = \tfrac{1}{2}\dfrac{0.005}{\sqrt{7.36}}$

$$\simeq \tfrac{1}{2}\dfrac{0.005}{2.7} \simeq 0.001$$

which is the same result as obtained on page 16.

Example. The numbers a and b when correctly rounded to three decimal places are 0·359 and 0·745 respectively. Use tables to obtain as accurate approximations as possible to $\cos a$ and $\cos b$.

From six-figure tables $\cos(0.359) \simeq 0.936249$.

Now since the derivative of $\cos x$ is $-\sin x$ and the maximum absolute error in 0·359 due to round-off is 0·0005, we see that an approximation to the corresponding maximum absolute error when $\cos(0.359)$ is used as an approximation for $\cos a$ is $0.0005|-\sin(0.359)| \simeq 0.0002$. (Note that there is no need for anything like six-figure accuracy in the value of $\sin(0.359)$.)

Hence $\cos a = 0.9362 \pm 0.0002$, or 0·936 correct to three decimal places.

Also, from tables $\cos(0.745) \simeq 0.735088$,

and $$0.0005|-\sin(0.745)| \simeq 0.0003.$$

Hence $\cos b = 0.7351 \pm 0.0003$, or 0·735 correct to three decimal places.

EXERCISES

1. The numbers 11·029, 2·3452, 13·374 and 0·00855 are correctly rounded approximations to the numbers a, b, c and d respectively.
   Evaluate approximations to
   (i) $a+b$
   (ii) $a+c$

      (iii) $a+d$
      (iv) $a-b$
    ✱ (v) $a-c$
      (vi) $a-d$
  ✱ (vii) $a+b-c$
    and discuss the error in each approximation.

2. A series of n numbers which have been correctly rounded, not all to the same number of decimal places, are to be added together to obtain an approximation to their sum. The least number of decimal places given in any of the numbers is d. Their sum can be approximated by
    (i) adding all the rounded numbers and then rounding their total to d decimal places, or
    (ii) rounding all the given numbers to d decimal places and then adding them.
Would you expect the results of (i) and (ii) to be identical? If not, which result would you expect to be the better approximation to the true sum?

3. With a, b, c and d as in exercise 1, evaluate approximations to
    (i) ac
    (ii) ad
  ✱ (iii) abc
  ✱ (iv) ab/c
    (v) $1/ac$
    and discuss the error in each approximation.

4. The numbers 2·14 and 8·27 are correctly rounded approximations to the numbers a and b respectively. Using tables of square roots, obtain as accurate approximations as possible to $\sqrt{a}$ and $\sqrt{b}$ and state the errors in these approximations.

5. The numbers a and b when correctly rounded to three decimal places are 0·359 and 0·745 respectively. Use tables to obtain as accurate approximations as possible to $\sin a$ and $\sin b$.

2.7. Reducing round-off error by choice of method

The effects of round-off errors can be very different using different numerical methods (or algorithms) to calculate the same quantity. This will be illustrated by considering a few specific examples.

Example. Consider the evaluation of $\sqrt{(x+1)}-\sqrt{x}$ for $x = 700$ and using four significant digits in the calculation.

$$\sqrt{701}-\sqrt{700} = 26\cdot48-26\cdot46$$
$$= 0\cdot02$$

Since both 26·48 and 26·46 have been rounded off, the error in the final result will lie between $\pm 0\cdot01$ so that we do not even have one-figure accuracy. The essential difficulty is that we have subtracted two "very nearly equal" numbers, and this has inevitably resulted in a loss of significant digits. One way of overcoming the difficulty is to use more significant digits in the intermediate calculations. This may not be convenient, however, for a variety of reasons. For example, four-figure tables of square roots are readily available, whereas seven or eight-figure tables are not so common. Thus, a better way of overcoming the difficulty is to use an alternative method to calculate the desired quantity. For example, we might proceed in the following way.

$$\sqrt{701} - \sqrt{700} = (\sqrt{701} - \sqrt{700})\left(\frac{\sqrt{701} + \sqrt{700}}{\sqrt{701} + \sqrt{700}}\right)$$

$$= \frac{1}{\sqrt{701} + \sqrt{700}}$$

$$= \frac{1}{52 \cdot 94}$$

$$= 0 \cdot 01889$$

The (approximate) relative error in this result is less than or equal to $\frac{0 \cdot 01}{52 \cdot 94} \simeq 0 \cdot 00019$ and so the (approximate) absolute error is less than or equal to $0 \cdot 000004$. Hence, using this method and again using only four significant digits in the calculations, we have obtained at least three significant digits in the final result.

Example. Evaluate the roots of the quadratic equation

$$x^2 - 60x + 1 = 0$$

using four significant digits throughout the calculations.

It is well known that the roots of the quadratic equation

$$ax^2 + bx + c = 0 \qquad (a \neq 0)$$

are given by

$$\frac{-b \pm \sqrt{(b^2 - 4ac)}}{2a}.$$

Therefore, in the particular case being considered, the roots are given by

$$\frac{60 \pm \sqrt{(60^2 - 4)}}{2} = 30 \pm \sqrt{899}$$

$$\simeq 30 \pm 29 \cdot 98$$

(where the square root is given correct to four significant digits).

Hence we obtain for the two roots $x_1 = 59 \cdot 98$ and $x_2 = 0 \cdot 02$. Thus while we have obtained one root (x_1) to four significant digits, the other root (x_2) has only been obtained to one significant digit. Starting from exact data (the coefficients of the polynomial) and working always to four significant digits accuracy, we have obtained only one significant digit in one of the roots. Once again, the essential difficulty is that to calculate the root x_2 involved the subtraction of two "nearly equal" numbers, thus resulting in the loss of significant digits. It is not difficult to construct a polynomial for which, again working to four significant digits and using the above formula, no significant digits at all are obtained for one of the roots (for example $x^2 - 220x + 1 = 0$).

To overcome the difficulty we can use a method very similar to that used in the previous example to calculate the smaller root x_2.

$$x_2 = 30 - \sqrt{899} = (30 - \sqrt{899}) \frac{30 + \sqrt{899}}{30 + \sqrt{899}}$$

$$= \frac{1}{30 + \sqrt{899}} \simeq \frac{1}{59 \cdot 98}$$

$$\simeq 0 \cdot 01667$$

(using four significant digits in the calculations).

The (approximate) absolute error in this result is less than or equal to $0 \cdot 00001$ so that at least three significant digits have been obtained.

Note that this method for calculating x_2 is equivalent to using

$$x_2 = \frac{\text{(product of the roots of the quadratic)}}{x_1}$$

$$= \frac{1}{30 + \sqrt{899}}$$

Thus, in general, when solving a quadratic equation the root with larger numerical value should be found first and the second root determined by dividing the product of the roots by this value.

EXERCISES

1. Evaluate $\sqrt{479} - \sqrt{478}$ correct to three significant figures assuming that only four significant figures can be retained in any calculations.

2. Evaluate both roots of the quadratic equation

$$x^2 - 18x + 1 = 0$$

as accurately as you can assuming that only three significant figures can be retained in any calculations.

3. Evaluate approximations to the roots of the quadratic equation

$$x^2 + 100x - 4 = 0.$$

Consider the value of the polynomial at the approximate roots and reconstruct the polynomial from the approximate roots.

4. How would you compute $\dfrac{1 - \cos x}{\sin x}$ for small values of x?

5. For values of x near 4, consider the computation of

$$\frac{\dfrac{1}{\sqrt{x}} - \dfrac{1}{2}}{x - 4}$$

Evaluate the expression at $x = 3\cdot9$.

2.8. Statistical treatment of errors

It must be emphasized that the analysis of this chapter has been concerned with obtaining upper limits for the absolute errors in the various calculations. Most often these upper limits will be considerably larger than the actual errors occurring, however, and the reason for this is not difficult to appreciate. For example, consider the addition of n numbers which have been rounded to k decimal places. The maximum absolute round-off error in each of the numbers is $\frac{1}{2}10^{-k}$ and so the maximum absolute error in their sum is $\frac{1}{2}n10^{-k}$. However, in order to obtain an error of such magnitude in the sum, the error in each individual number would have to have had its maximum possible magnitude and each of these errors would also have to have been of the same sign. Clearly the chances of this occurring in practice are very small indeed. Normally the magnitude of the individual errors will vary and some will be positive and some negative (that is some numbers will be rounded up while others will be rounded down) so that some cancellation of errors will occur. Assuming a random distribution of errors it can be shown using statistical theory that for large n a more realistic estimate of the error occurring in the sum of n numbers each rounded to k decimal places is $\frac{1}{2}\sqrt{n}10^{-k}$. Similar results can of course be obtained for the effects of errors on the other basic operations of arithmetic.

It must however be emphasized that these statistical results do not apply when only a few numbers are being operated on. Further, in any particular case, we can be 100% sure only that the errors involved lie within the upper limits obtained in the earlier sections.

Evaluation of formulae

IF THE VALUES OF A FUNCTION f ARE REQUIRED FOR VALUES OF THE independent variable at intervals of p starting at a and finishing at b, where it is implied that $(b-a)$ is equal to a positive integer times p, then we write "evaluate f for $a(p)b$".

For example, evaluate x^2 for 1(1)10 means determine the values of x^2 at unit intervals of x from $x = 1$ to $x = 10$, that is for $x = 1, 2, 3, 4, 5, 6, 7, 8, 9, 10$.

The following table shows the values, correct to three decimal places, of f for 1(0·1)2 where $f(x) \equiv 1/x$.

x	$f(x)$
1	1·000
1·1	·909
1·2	·833
1·3	·769
1·4	·714
1·5	·667
1·6	·625
1·7	·588
1·8	·556
1·9	·526
2	·500

A flow diagram for evaluating $f(x)$ for $a(p)b$ is shown opposite.

EXERCISES

1. Tabulate $\dfrac{x}{x^2+1}$ for 0(1)10 correct to four decimal places.
2. Tabulate $\log(x^{\frac{1}{2}}+1)$ for 5(5)25.
3. Tabulate $3 \sin x°$ for 0°(10°)90°.

3.1. The evaluation of polynomials (nesting)

Consider the evaluation of the polynomial $2x^3 - 5x^2 - 3x + 4$ when $x = \alpha$. One way of doing this is first to calculate α^2 and α^3 ($= \alpha \times \alpha^2$). This would

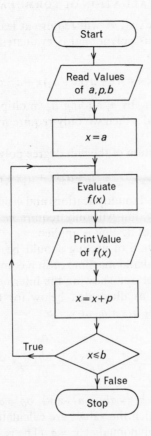

```
        ( Start )
           │
           ▼
    / Read Values /
    /  of a,p,b  /
           │
           ▼
    ┌───────────┐
    │   x = a   │
    └───────────┘
           │
           ▼
    ┌───────────┐
    │ Evaluate  │
    │   f(x)    │
    └───────────┘
           │
           ▼
   / Print Value /
   /  of f(x)   /
           │
           ▼
    ┌───────────┐
    │ x = x + p │
    └───────────┘
           │
           ▼
  True   ◇       ◇
  ◄──────  x ≤ b
           ◇       ◇
           │ False
           ▼
        ( Stop )
```

involve two multiplications. To complete the evaluation would then require three more multiplications (one corresponding to each of the coefficients of x, x^2 and x^3) and three additions. Thus the whole process, carried out in this way, would require a total of five multiplications and three additions.

Now suppose that we rewrite the given polynomial in the form

$$((2x - 5)x - 3)x + 4$$

and evaluate it at $x = \alpha$ in the way suggested by this bracketed form. Carried out in this way then the whole evaluation requires only three multiplications and three additions. This method of evaluation is known as *nesting*.

Similarly, evaluation of the polynomial

$$a_5 x^5 + a_4 x^4 + a_3 x^3 + a_2 x^2 + a_1 x + a_0$$

25

by first calculating α^2, α^3, α^4, α^5 will require at least nine multiplications and five additions. If this polynomial is evaluated in the way suggested by writing it in the form

$$((((a_5x + a_4)x + a_3)x + a_2)x + a_1)x + a_0$$

that is by multiplying a_5 by α, adding a_4, multiplying by α, adding a_3, etc.—the nesting method—then we only require five multiplications and five additions.

In general, the evaluation of the mth degree polynomial

$$p(x) = a_m x^m + a_{m-1} x^{m-1} + \cdots + a_2 x^2 + a_1 x + a_0 \quad \text{with } a_m \neq 0$$

will require at least $(2m-1)$ multiplications and m additions if $\alpha^2, \alpha^3 \ldots, \alpha^m$ are evaluated separately, but will only require m multiplications and m additions if the method of nesting is applied.

Clearly then the method of nesting should be used when evaluating polynomials either on a hand machine or in a computer.

The process can be set out, showing the intermediate values obtained during the calculation, as illustrated below for the evaluation of the polynomial $a_3 x^3 + a_2 x^2 + a_1 x + a_0$ at $x = \alpha$.

α	a_3	a_2	a_1	a_0
	$\downarrow$ $\nearrow$	αb_3 $\nearrow$	αb_2 $\nearrow$	αb_1 $\nearrow$
	b_3	b_2	b_1	b_0

$b_3 = a_3$, $b_2 = \alpha b_3 + a_2$, $b_1 = \alpha b_2 + a_1$ and $b_0 = \alpha b_1 + a_0$. The arrows indicate the order in which the values are calculated. The value of b_0 is the value of the given polynomial at $x = \alpha$. (The reader should verify this for himself.)

If a hand machine is being used it is not, of course, necessary to record any intermediate values.

Example. Evaluate the cubic $2x^3 - 5x^2 - 3x + 4$ when $x = 0.5$.

0.5	2	-5	-3	4
		1	-2	-2.5
	2	-4	-5	1.5

Hence the value of the given cubic at 0.5 is 1.5.

Example. Evaluate the quartic $3x^4 + 5x^3 - x^2 - 2x + 1$ when $x = -1.2$.

-1.2	3	5	-1	-2	1
		-3.6	-1.68	3.216	-1.4592
	3	1.4	-2.68	1.216	-0.4592

Hence the value of the given quartic at $-1\cdot2$ is $-0\cdot4592$.

3.1.1. Flow Diagram

A flow diagram for the evaluation of the mth degree polynomial $a_m x^m + a_{m-1} x^{m-1} + \cdots + a_1 x + a_0$ is shown below.

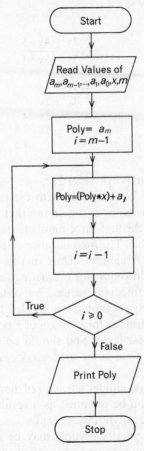

EXERCISES

1. Evaluate the polynomial

$$2x^3 - 3x^2 + x - 4$$

at $x = 1\cdot75$.

2. Evaluate the polynomial

$$3x^4 - x^3 - 2x - 5$$

at $x = 0\cdot683$. Give the answer correct to three decimal places.

3.2. Synthetic division of a polynomial by a linear factor

In this section we consider a method of obtaining the quotient polynomial and the remainder when a polynomial is divided by a linear factor.

For example, let us consider the long division of the cubic polynomial $2x^3 - 5x^2 - 3x + 4$ by the linear factor $(x - 0.5)$.

This proceeds as follows:

$$
\begin{array}{r}
2x^2 - 4x - 5 \\
x - 0.5\overline{)2x^3 - 5x^2 - 3x + 4} \\
\underline{2x^3 - x^2} \\
-4x^2 - 3x \\
\underline{-4x^2 + 2x} \\
-5x + 4 \\
\underline{-5x + 2.5} \\
1.5
\end{array}
$$

Thus the quotient is the polynomial $2x^2 - 4x - 5$ and the remainder is 1·5. These results should be compared with the table drawn up for the example on page 26. This comparison shows that the coefficients of the quotient polynomial are the first three numbers (starting from the left) in the bottom row of the table. The final number on the right of the bottom row of the table is the remainder. Thus in this case we see that, by carrying out the nesting procedure for evaluating the given polynomial at 0·5 and noting the intermediate values, we obtain the quotient and remainder when the given polynomial is divided by the factor $(x - 0.5)$. This result is true in general for the division of a polynomial of arbitrary (integral) degree by a linear factor and should be verified by the reader for the division of the general cubic $a_3x^3 + a_2x^2 + a_1x + a_0$ by the linear factor $(x - \alpha)$.

Thus the quotient and remainder when a polynomial $p(x)$ is divided by the linear factor $(x - \alpha)$ can be obtained as a result of using the nesting procedure to evaluate the polynomial $p(x)$ at α.

The process, being essentially nesting, may be recorded in the same tabular form as the nesting process and is called *synthetic division*.

Example. Use the method of synthetic division to obtain the quotient and remainder when the polynomial $3x^5 + 5x^4 + 8x^2 + 7x + 4$ is divided by $(x + 2)$.

We construct the table for the evaluation of the given polynomial at -2 using the nesting process.

This is

$$
\begin{array}{r|rrrrrr}
-2 & 3 & 5 & 0 & 8 & 7 & 4 \\
   &   & -6 & 2 & -4 & -8 & 2 \\
\hline
   & 3 & -1 & 2 & 4 & -1 & 6 \\
\end{array}
$$

Hence the quotient is $3x^4 - x^3 + 2x^2 + 4x - 1$ and the remainder is 6.

EXERCISES

1. Obtain the quotient and remainder when the polynomial

$$5x^4 + 3x^3 - x^2 + 6x - 7$$

is divided by $x - 2$.

2. Obtain the quotient and remainder when the polynomial

$$3x^7 + x^6 - 4x^5 + 2x^4 + 2x^3 - x^2 + x - 1$$

is divided by $x + 3$.

3. Obtain the quotient and remainder when the polynomial

$$2 \cdot 31792x^4 + 6 \cdot 17843x^3 - 4 \cdot 40632x^2 + 5 \cdot 16931x - 9 \cdot 76802$$

is divided by $x + 3 \cdot 47918$. (Use six significant figures in the calculations.)

4. Obtain the quotient and remainder when the polynomial

$$3 \cdot 40617x^5 - 5 \cdot 68742x^4 - 1 \cdot 02795x^2 + 2 \cdot 30469$$

is divided by $x - 0 \cdot 629734$. (Use six significant figures in the calculations.)

5. Express the quotient polynomial and the remainder when a polynomial $p(x)$ is divided by $(ax - b)$, in terms of the quotient polynomial and the remainder when $p(x)$ is divided by $(x - b/a)$. Hence use the method of synthetic division to determine the quotient and remainder when the polynomial

$$x^4 - 2x^3 - 11x^2 + 12x + 36$$

is divided by $2x + 1$.

Finite differences

4.1. Finite difference tables

The distance s travelled by a car from a given fixed point will depend on the time t for which the car has been travelling. Indeed, since for any particular time the car must have travelled a unique distance, the distance is a function of time. Let $s = f(t)$. Now suppose that when the distance travelled (s metres) is measured at successive 10-second intervals in time (t seconds) we obtain the following results:

t	$s = f(t)$
0	0
10	214
20	736
30	1446
40	2270
50	3164
60	4100

In order to extract more information from the above data it is often useful to rewrite the table in the expanded form of a *finite-difference table* as shown:

t	$f(t)$	1st Differences	2nd Differences	3rd Differences	4th Differences
0	0				
		214			
10	214		308		
		522		−120	
20	736		188		46
		710		−74	
30	1446		114		30
		824		−44	
40	2270		70		16
		894		−28	
50	3164		42		
		936			
60	4100				

The first differences are obtained by subtracting each function value from the one immediately below it in the table. For example, the top two entries in the column of first differences are obtained by subtracting 0 from 214 and 214 from 736. The second differences are obtained in the same way, that is, by subtracting each first difference from the one immediately below it in the table. For example, the top entry in the column of second differences is obtained by subtracting 214 from 522. Similarly third and higher-order differences can be obtained. It is customary to write even-order differences on the same lines as the function values, while odd-order differences are written on intermediate lines (see the above table). There are three different notations for differences in common use. These different notations make use of the symbols δ, Δ and ∇ which are all pronounced "delta".

4.1.1. Central differences

The first notation is that for central differences. Here we denote the first difference $f(t_1) - f(t_0)$ by $\delta f_{1/2}$. The first difference $f(t_2) - f(t_1)$ is denoted by $\delta f_{3/2}$ and so on. In general $(f_{i+1/2} - f_{i-1/2}) = \delta f_i$. Similarly the second difference $(\delta f_{3/2} - \delta f_{1/2})$ is denoted by $\delta^2 f_1$, etc. Thus, in central-difference notation, the difference table for a function f of an independent variable t has the form

t_0	f_0			
t_1	f_1	$\delta f_{1/2}$		
		$\delta f_{3/2}$	$\delta^2 f_1$	
t_2	f_2	$\delta f_{5/2}$	$\delta^2 f_2$	$\delta^3 f_{3/2}$
t_3	f_3	$\delta f_{7/2}$	$\delta^2 f_3$	$\delta^3 f_{5/2}$
t_4	f_4			

in which $f_0 = f(t_0)$, $f_1 = f(t_1), \ldots, f_4 = f(t_4)$.

Example. Evaluate the polynomial $f(x) \equiv x^3 - 8x^2 - 4x + 1$ for integer values of x between $x = 1$ and $x = 10$ and construct the difference table.

x	$f(x)$				
0	1				
1	−10	−11			
		−21	−10		
2	−31	−25	−4	6	
3	−56	−23	2	6	0
4	−79	−15	8	6	0
5	−94	−1	14	6	0
6	−95	19	20	6	0
7	−76	45	26	6	0
8	−31	77	32	6	0
9	46	115	38	6	
10	161				

Note that the third differences all have the same value (namely 6) and so all fourth and higher differences will be zero.

It is customary to omit decimal points and to use only significant figures when writing down differences in a table. When individual differences are quoted, however, they should be given in full with the decimal point shown (if appropriate).

Example. Evaluate the polynomial $f(x) \equiv x^3 - 8x^2 - 4x + 1$ for $x = 0(0{\cdot}1)0{\cdot}5$ and construct the difference table.

x	$f(x)$				
0	1				
		-479			
0·1	0·521		-154		
		-633		6	
0·2	$-0{\cdot}112$		-148		0
		-781		6	
0·3	$-0{\cdot}893$		-142		0
		-923		6	
0·4	$-1{\cdot}816$		-136		
		-1059			
0·5	$-2{\cdot}875$				

Note that the third differences all have the value 0·006.

It is not necessary to use t_0 at the top of the table; instead t_0 may, for example, be used to denote a value of t somewhere in the middle of the interval. A difference table in central-difference notation with t_0 denoting the middle value has the form

t_{-2}	f_{-2}				
		$\delta f_{-3/2}$			
t_{-1}	f_{-1}		$\delta^2 f_{-1}$		
		$\delta f_{-1/2}$		$\delta^3 f_{-1/2}$	
t_0	f_0		$\delta^2 f_0$		
		$\delta f_{1/2}$		$\delta^3 f_{1/2}$	
t_1	f_1		$\delta^2 f_1$		
		$\delta f_{3/2}$			
t_2	f_2				

Referring to the numerical table on page 30, if $t_0 = 30$ then $\delta f_{-3/2} = 522$, $\delta^2 f_1 = 70$, etc.

4.1.2. Forward differences

In forward-difference notation we denote the first difference $f(t_1) - f(t_0)$ by Δf_0, the first difference $f(t_2) - f(t_1)$ by Δf_1, the second difference $(\Delta f_1 - \Delta f_0)$ by $\Delta^2 f_0$, etc., so that (in this notation) the above difference table would have the form

t_0	f_0			
		Δf_0		
t_1	f_1		$\Delta^2 f_0$	
		Δf_1		$\Delta^3 f_0$
t_2	f_2		$\Delta^2 f_1$	
		Δf_2		$\Delta^3 f_1$
t_3	f_3		$\Delta^2 f_2$	
		Δf_3		
t_4	f_4			

As for central differences it is not necessary to use t_0 at the top of the table.

4.1.3. Backward differences

In backward-difference notation we denote the first difference $f(t_1) - f(t_0)$ by ∇f_1, the first difference $f(t_2) - f(t_1)$ by ∇f_2, the second difference $(\nabla f_2 - \nabla f_1)$ by $\nabla^2 f_2$, etc., so that (in this notation) the above difference table would have the form

$$
\begin{array}{lllll}
t_0 & f_0 & & & \\
t_1 & f_1 & \nabla f_1 & \nabla^2 f_2 & \\
t_2 & f_2 & \nabla f_2 & \nabla^2 f_3 & \nabla^3 f_3 \\
t_3 & f_3 & \nabla f_3 & \nabla^2 f_4 & \nabla^3 f_4 \\
t_4 & f_4 & \nabla f_4 & &
\end{array}
$$

As for central and forward differences it is not necessary to use t_0 at the top of the table.

4.1.4. The relationship between the different notations

It is important to note that, despite the differences in notation, in any particular example it is the *same numbers* which will appear in the same positions in all three tables; that is, $\delta f_{1/2}$, Δf_0, ∇f_1 are simply three different ways of denoting the *same* number. Similarly $\delta^2 f_1$, $\Delta^2 f_0$, $\nabla^2 f_2$ represent the *same* number, and so on.

From the example on page 32, we see that if $x_0 = 0.3$, then

$$\delta f_{1/2} = \Delta f_0 = \nabla f_1 = -0.923,$$
$$\delta f_{-5/2} = \Delta f_{-3} = \nabla f_{-2} = -0.479,$$
$$\delta^2 f_{-2} = \Delta^2 f_{-3} = \nabla^2 f_{-1} = -0.154,$$
$$\delta^2 f_1 = \Delta^2 f_0 = \nabla^2 f_2 = -0.136,$$

etc.

EXERCISE

Evaluate $f(x) \equiv 1/x$ correct to 4 decimal places for $x = 1(1)10$ and construct the difference table as far as fourth differences.
With $x_0 = 5$, what are the values of $\delta f_{-1/2}$, $\delta^2 f_3$, $\delta^4 f_{-2}$, Δf_2, $\Delta^2 f_3$, $\Delta^3 f_{-3}$, ∇f_3, $\nabla^3 f_{-1}$, $\nabla^4 f_4$?

4.2. Propagation of round-off errors in difference tables

Suppose now that our function values are not known exactly but are subject to round-off errors. If the rounding is to 2 decimal places, the maximum numerical error in each function value is 0·005. Then the maximum numerical error in each first difference is 0·01, that is, one unit in the second decimal place. The maximum numerical error in each second difference is 0·02 or two units in the second decimal place. The maximum numerical error in each third difference is 0·04 or 4 units in the second decimal place and so on. Similar results, of course, hold when the function values are rounded to k decimal places for any integer k.

In the following difference table the function values have been obtained correct to two decimal places:

x	$f(x)$			
1	−41·23			
2	−39·89	134		
		492	358	
3	−34·97	848	356	−2
4	−26·49	1206	358	2
5	−14·43	1561	355	−3
6	1·18	1919	358	3
7	20·37	2278	359	1
8	43·15			

The third differences here are all approximately zero to within the accuracy ($\pm 0·04$) permitted by round-off errors in the function values. The second differences are all approximately constant (namely 3·57) to the accuracy ($\pm 0·02$) permitted by round-off errors in the function values.

4.3. Differences of polynomials

We begin this section by constructing the difference table for the polynomial $f(x) \equiv x^2 + 2x - 1$ for (exact) values of $x = 1·00(0·2)1·10$.

x	$f(x)$		
1·00	2		
1·02	2·0804	804	8
1·04	2·1616	812	8
1·06	2·2436	820	8
1·08	2·3264	828	8
1·10	2·4100	836	

The important point to note is that for this *second-degree polynomial* the *second differences are constant*.

We recall too that in the example on page 31 we saw that for the third-degree polynomial considered there the third differences were constant.

These two examples are illustrations of the general theorem which states that the nth differences of the nth degree polynomial

$$a_n x^n + a_{n-1} x^{n-1} + \cdots + a_1 x + a_0 \quad (a_n \neq 0)$$

are $n!a_n h^n$ where h is the difference between successive values of x at which the polynomial is evaluated. The proof of this theorem is by induction and is given in Appendix 2.

It is important to point out that this theorem applies only when the function values are exact. If round-off errors are present, the result will not generally hold. However, it will be true for the general nth degree polynomial that its nth differences will have the constant value $n!a_n h^n$ to within the limits imposed by round-off errors (see §4.2).

For example, consider the following difference table for the cubic $x^3 - 8x + 5$. The values of the cubic are rounded to two decimal places.

x	$f(x)$			
2·0	−3·00			
		105		
2·2	−1·95		52	
		157		7
2·4	−0·38		59	
		216		2
2·6	1·78		61	
		277		7
2·8	4·55		68	
		345		4
3·0	8·00		72	
		417		4
3·2	12·17		76	
		493		
3·4	17·10			

Clearly the third differences here do not have the constant value $3!(\cdot 2)^3 = 0\cdot048$. However they are all in fact approximately equal to this value to within the limits ($\pm 0\cdot04$) imposed by the possible round-off errors in the polynomial values.

EXERCISE

Verify the result of the theorem of this section for the following cases:
  (i) the polynomial $2x^2 - x + 1$ and $x = 0(1)5$
 (ii) the polynomial $3x^3 + x^2 - 4x - 5$ and $x = -2(0\cdot5)2$
(iii) the polynomial $-x^4 + x^2 + 3x + 1$ for $x = 1(0\cdot1)1\cdot5$.

4.4 Fitting a polynomial to given function values

The process to be considered in this section is best illustrated by an example. Given a set of values of a function $f(x)$ corresponding to a set of equally spaced values of x, determine the polynomial of lowest degree which fits them.

x	0	0·5	1·0	1·5	2·0	2·5	3·0
$f(x)$	0·25	−1·50	−1·75	−0·50	2·25	6·50	12·25

The first step is to construct the difference table shown below.

x	$f(x)$		
0	0·25		
		−175	
0·5	−1·50		150
		−25	
1·0	−1·75		150
		125	
1·5	−0·50		150
		275	
2·0	2·25		150
		425	
2·5	6·50		150
		575	
3·0	12·25		

Since the second differences are constant, the required polynomial must have degree two. Further, we must have $2!a_2(0·5)^2 = 1·50$ where a_2 is the coefficient of x^2 in the required polynomial. Therefore $a_2 = 3$.

Now form the difference table for the function $g(x) = f(x) - 3x^2$. This function must be able to be fitted at the given points by a polynomial of first degree, and so its first differences should be constant.

x	$g(x)$	
0	0·25	
		−250
0·5	−2·25	
		−250
1·0	−4·75	
		−250
1·5	−7·25	
		−250
2·0	−9·75	
		−250
2·5	−12·25	
		−250
3·0	−14·75	

Then $1!a_1(0·5) = -2·50$ where a_1 is the coefficient of x in the required polynomial. Therefore $a_1 = -5$.

Now the values of $f(x) - 3x^2 + 5x$ are all equal to 0·25, so that the required polynomial is $3x^2 - 5x + 0·25$.

In this case the given function values could be fitted exactly by the quadratic. The method can, however, be extended to the case in which the given data values can be fitted only approximately by a polynomial of low degree. An alternative method for finding a polynomial fit to given data is given in the next chapter in §5.3.

It must be emphasized most strongly that, once a polynomial has been obtained which fits the values of a function for a given set of values of the independent variable, it cannot be assumed (without more information) that it will also fit the function at other values of the independent variable. This fact is also best illustrated by an example. The function $\cos 2\pi x$ evaluated at $x = 0, 1, 2, \ldots$ has the constant value 1. Hence, if we are given as data the values of this function at these values of x, they can be fitted by the polynomial $p(x) \equiv 1$. Clearly the polynomial does not fit the values of $\cos 2\pi x$ at $x = \frac{1}{2}, \frac{3}{2}, \frac{5}{2}, \ldots$ The problem of determining when a polynomial which fits given function values for specific values of the independent variable also fits the function values at intermediate values of the independent variable is an interesting one which is unfortunately beyond the scope of this book.

EXERCISE

Evaluate the function $f(x) \equiv 3x^2 - 5x + 0.25 + 6 \sin 2\pi x$ for $x = 0(0.5)3.0$ and determine the polynomial which fits these function values. Comment on your answer.

4.5. Relationship between differences and derivatives

The derivative of a function f at x_0 can be expressed as

$$f'(x_0) = \lim_{h \to 0} \frac{f(x_0 + \frac{1}{2}h) - f(x_0 - \frac{1}{2}h)}{h} = \lim_{h \to 0} \frac{1}{h} \delta f_0.$$

Hence, for sufficiently small h, we have

$$f'(x_0) \simeq \frac{1}{h} \delta f_0.$$

However, the values $f(x_0 + \frac{1}{2}h)$ and $f(x_0 - \frac{1}{2}h)$ are not tabulated in a difference table but instead we have the values $f(x_0)$ and $f(x_0 + h)$. These values can be used to give an approximation to $f'(x_0 + \frac{1}{2}h)$. We have

$$f'(x_0 + \tfrac{1}{2}h) \simeq \frac{1}{h} \delta f_{1/2} = \frac{1}{h} \{ f(x_0 + h) - f(x_0) \}$$

or

$$f'_{1/2} \simeq \frac{1}{h} \delta f_{1/2}.$$

Now

$$f''(x_0) = \lim_{h \to 0} \frac{f'(x_0 + \frac{1}{2}h) - f'(x_0 - \frac{1}{2}h)}{h}.$$

Hence, for sufficiently small values of h,

$$f''(x_0) \simeq \frac{1}{h} \{ f'(x_0 + \tfrac{1}{2}h) - f'(x_0 - \tfrac{1}{2}h) \}$$

$$\simeq \frac{1}{h} \{ \frac{1}{h} \delta f_{1/2} - \frac{1}{h} \delta f_{-1/2} \}$$

$$= \frac{1}{h^2} \delta^2 f_0.$$

These results are special cases of the general result, which can be proved by induction, that

$$f_0^{(n)} \simeq \frac{1}{h^n} \delta^n f_0$$

in which $f_0^{(n)}$ denotes the nth derivative of the function f evaluated at x_0.

It is interesting to note how the central result of this section ties up with the central result of §4.3. In the latter section it was stated that the nth differences of an nth degree polynomial in which the coefficient of x^n is a_n have the constant value $n!a_n h^n$. Now, since for positive integers $m < n$ (and for $m = 0$) the nth derivative of x^m is zero and the nth derivative of x^n is $n!$, we see that the nth derivative of the nth degree polynomial $p_n(x) \equiv a_n x^n + a_{n-1}x^{n-1} + \cdots + a_1 x + a_0$ is $n!a_n$. Hence we have

$$p_n^{(n)}(x) = \frac{1}{h^n} \delta^n p_n(x).$$

The above approximate formulae are examples of numerical differentiation. Numerical differentiation is not, however, a very satisfactory process. To illustrate this we will consider the process for the first derivative.

We have

$$f_{1/2}' \simeq \frac{1}{h} \delta f_{1/2} = \frac{f(x_0 + \frac{1}{2}h) - f(x_0 - \frac{1}{2}h)}{h}$$

Now the truncation error of this formula is decreased by taking smaller and smaller values of h. However, decreasing h will have the effect of making the function values $f(x_0 + \frac{1}{2}h)$ and $f(x_0 - \frac{1}{2}h)$ more nearly equal, so that we are faced with the problem of loss of significant figures associated with subtracting two nearly equal numbers.

Example. Obtain a numerical approximation to the second derivative of the logarithmic function at 5 using four-figure logarithm tables and (i) $h = 0.5$, (ii) $h = 0.25$, (iii) $h = 0.125$. (To three significant figures the answer is -0.0174.)

We shall use the result

$$f''(x_0) \simeq \frac{1}{h^2} \delta^2 f_0.$$

(i) We require the second central difference of the logarithmic function at 5. To obtain this with $h = 0.5$ we require only the values of the logarithm at 4.5, 5, 5.5.

x	$\log x$		
4·5	0·6532		
5·0	0·6990	458	−44
5·5	0·7404	414	

Hence
$$f''(5) \simeq \frac{1}{(0·5)^2}(-0·0044) = -0·0176.$$

(ii) As above, we require the values of the logarithm at 4·75, 5, 5·25.

x	$\log x$		
4·75	0·6767		
5·00	0·6990	223	−11
5·25	0·7202	212	

Hence
$$f''(5) \simeq \frac{1}{(0·25)^2}(-0·0011) = -0·0176.$$

Thus decreasing h has not changed the accuracy of the result.
(iii) As above, we require the values of the logarithm at 4·875, 5, 5·125.

x	$\log x$		
4·875	0·6879		
5·000	0·6990	111	−4
5·125	0·7097	107	

Hence
$$f''(5) \simeq \frac{1}{(0·125)^2}(-0·0004) = -0·0256.$$

In this case, decreasing h has resulted in a much poorer answer.

Interpolation

SUPPOSE WE ARE GIVEN THE VALUES OF A FUNCTION f CORRESPONDING
to several values $x_0, x_1, x_2, \ldots, x_n$ of the independent variable x. Interpolation is then the process by which we obtain, from these function values,
approximations to the function values for values of the independent
variable which are intermediate to those given above, say for example,
at a value of x between x_4 and x_5.

For example, we may be given the following information

x	$f(x) \equiv \log x$
80	1·9031
81	1·9085
82	1·9138
83	1·9191
84	1·9243
85	1·9294

and asked to determine approximations to $\log 80\cdot 5$, $\log 82\cdot 75$ or $\log 84\cdot 8$
etc., or given the information

x	$f(x) \equiv e^x$
0·33	1·3910
0·34	1·4049

and asked to estimate $e^{0\cdot 333}$.

One way of proceeding is to attempt to draw the graph of the curve
represented by $y = f(x)$. We can then read off the value of the function
f corresponding to a particular value of x. This is not as easy as it may
appear, however, especially if we are simply given several function values
and no further information about the function. The problem is seen to be
even more difficult when we remember that the function values themselves
are almost certainly not exact. They have most likely been rounded off
to a given accuracy, even if they are not subject to other errors such as
experimental errors.

For example, if we are simply given the data

x	$f(x)$
x_0	$f(x_0)$
x_1	$f(x_1)$
x_2	$f(x_2)$

without any further information on the function f, we can readily plot
the points $(x_0, f(x_0))$, $(x_1, f(x_1))$ and $(x_2, f(x_2))$ on graph paper as in
Figure 1a. However we have no way of telling whether the graph of
$y = f(x)$ looks like that shown in Figure 1b or like that in Figure 1c or
indeed is something which looks different from either of those.

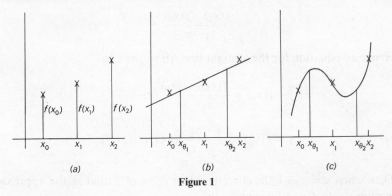

(a)   (b)   (c)

Figure 1

Clearly the approximations obtained for the function values at $x = x_{\theta_1}$
and $x = x_{\theta_2}$ from Figure 1b are very different from the approximations
obtained from Figure 1c.

5.1. Linear interpolation—
Newton's forward-difference formula of degree one

Suppose we are given the function values $Y_0 = f(x_0)$, $Y_1 = f(x_1)(x_1 > x_0)$
and that we desire to obtain an approximation to $f(x_0 + k)$ where
$x_0 < x_0 + k < x_1$. Clearly if we knew the shape of the curve $y = f(x)$
between $x = x_0$ and $x = x_1$ we could determine $f(x_0 + k)$ very accurately.
However, if the shape is not known, we can take as a first approximation
a straight line, that is, we can assume that the curve is approximately linear
between the points $A(x_0, Y_0)$ and $B(x_1, Y_1)$ (Figure 2). Then an approximation to $Y_k = f(x_0 + k)$ can be obtained from the graph (Figure 2). It is

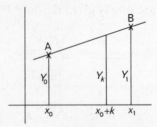

Figure 2

not necessary, however, to use the graph to determine the approximation to $f(x_0 + k)$; instead we can determine it analytically as follows.

The gradient of the straight line AB is

$$\frac{f(x_1) - f(x_0)}{x_1 - x_0}.$$

Hence an equation for the straight line AB is given by

$$y - f(x_0) = \frac{f(x_1) - f(x_0)}{x_1 - x_0} (x - x_0)$$

or

$$y = f(x_0) + \left(\frac{x - x_0}{x_1 - x_0}\right)(f(x_1) - f(x_0)).$$

Then when $x = x_0 + k$ the corresponding value of y (that is, the approximation $y_{k/h}$ to $f(x_0 + k)$) is given by

$$y_{k/h} = f(x_0) + \frac{k}{x_1 - x_0} (f(x_1) - f(x_0))$$

$$= f(x_0) + \frac{k}{h} (f(x_1) - f(x_0))$$

$$= f_0 + \frac{k}{h} \Delta f_0$$

where $x_1 - x_0 = h$.

If now the fraction k/h is denoted by θ, we have

$$y_{k/h} = y_\theta = (1 - \theta)f(x_0) + \theta f(x_1) = f_0 + \theta \Delta f_0.$$

In most applications θ will be between 0 and 1.

This method of determining an approximation to $f(x_0 + k)$ by assuming that the curve $y = f(x)$ is linear between $(x_0, f(x_0))$ and $(x_1, f(x_1))$ is called *linear interpolation*. The approximate formula

$$f(x_0 + \theta h) \simeq f_0 + \theta \Delta f_0$$

is called Newton's forward-difference interpolation formula of degree one.

Example. From the data given on page 40 determine approximations to log 80·5, log 82·75 and log 84·8 using linear interpolation.

x	$\log x$	*1st differences*
80	1·9031	
		54
81	1·9085	
		53
82	1·9138	
		53
83	1·9191	
		52
84	1·9243	
		51
85	1·9294	

Using Newton's forward-difference formula we have

$$\log 80·5 \simeq 1·9031 + \frac{0·5}{1}(0·0054)$$

$$= 1·9058.$$

(From 4-figure tables, log 80·5 = 1·9058.)

$$\log 82·75 \simeq 1·9138 + 0·75\,(0·0053)$$

$$= 1·9178 \text{ (to 4 decimal places)}.$$

(From 4-figure tables, log 82·75 = 1·9178.)

$$\log 84·8 \simeq 1·9243 + 0·8\,(0·0051)$$

$$= 1·9284 \text{ (to 4 decimal places)}.$$

(From 4-figure tables, log 84·8 = 1·9284.)

Example. From the data given on page 40, determine an approximation to $e^{0·333}$ using linear interpolation.

x	e^x	*1st differences*
0·33	1·3910	
		139
0·34	1·4049	

Then $\qquad e^{0·333} \simeq 1·3910 + \dfrac{0·003}{0·01}(0·0139) = 1·3952$ (to 4 decimal places).

(From tables $e^{0·333} = 1·3951$ to 4 decimal places.)

We must now discuss the errors in the application of linear interpolation. We shall not consider "mistakes" but shall look at the effect on the interpolated value of round-off (or other) errors in the given data and the error of the method itself (truncation error).

5.1.1. *Effect of round-off in the given data*

Let the given data be

x	$f(x)$
x_0	y_0
x_1	y_1

We shall assume that y_0 and y_1 are subject to round-off errors e_0 and e_1 respectively, so that the corresponding exact (but unknown) values Y_0 and Y_1 are given by

$$Y_0 = y_0 + e_0$$

and

$$Y_1 = y_1 + e_1.$$

Now the approximation y_θ to $f(x_0 + \theta h)$ which we calculate using linear interpolation is given by

$$y_\theta = (1 - \theta)y_0 + \theta y_1.$$

However, what we really want is Y_θ where

$$Y_\theta = (1 - \theta)Y_0 + \theta Y_1.$$

(Remember that we cannot actually calculate Y_θ as Y_0 and Y_1 are unknown, so the best we can do is to calculate y_θ.)

Hence the error e_θ in our approximation, due to the round-off errors in the given data, is given by

$$\begin{aligned} e_\theta &= Y_\theta - y_\theta \\ &= (1 - \theta)(Y_0 - y_0) + \theta(Y_1 - y_1) \\ &= (1 - \theta)e_0 + \theta e_1. \end{aligned}$$

Hence

$$|e_\theta| \leqslant (1 - \theta)|e_0| + \theta|e_1|$$

since $(1 - \theta)$ and θ are both positive for $0 < \theta < 1$.

Then, if both $|e_0|$ and $|e_1|$ are less than or equal to $e(> 0)$,

$$|e_\theta| \leqslant e.$$

For example, if both y_0 and y_1 are rounded to k decimal places then $|e_0|$ and $|e_1|$ are each less than or equal to $\frac{1}{2}10^{-k}$ and so $|e_\theta|$ will also be less than or equal to $\frac{1}{2}10^{-k}$; that is, in the absence of any truncation error of the method, the interpolated value will also be correct to k decimal places.

5.1.2. Truncation error of the method

This error is due to the fact that we have approximated to the curve between the points (x_0, y_0) and (x_1, y_1) by a straight line when in fact the curve has some other shape in that interval. It will in general be rather difficult to estimate this error accurately for given data.

Here we start from the data

x	$f(x)$
x_0	$f(x_0) = Y_0$
x_1	$f(x_1) = Y_1$

that is, we assume that the function values Y_0 and Y_1 are known exactly. The approximation to $f(x_0 + \theta h)$ which we obtain using linear interpolation is then

$$(1 - \theta) Y_0 + \theta Y_1.$$

Hence the truncation error is

$$f(x_0 + \theta h) - (1 - \theta) Y_0 - \theta Y_1 = f(x_0 + \theta h) - (1 - \theta) f(x_0) - \theta f(x_1).$$

But from Taylor's series expansion* we obtain

$$f(x_0 + \theta h) = f(x_0) + \theta h f'(x_0) + \tfrac{1}{2}(\theta h)^2 f''(x_0) + \cdots$$

and

$$f(x_1) = f(x_0 + h)$$
$$= f(x_0) + h f'(x_0) + \tfrac{1}{2} h^2 f''(x_0) + \cdots$$

Therefore the truncation error is

$$f(x_0) + \theta h f'(x_0) + \tfrac{1}{2}(\theta h)^2 f''(x_0) + \cdots$$
$$- f(x_0) + \theta f(x_0)$$
$$- \theta f(x_0) - \theta h f'(x_0) - \tfrac{1}{2}\theta h^2 f''(x_0) + \ldots$$
$$= \tfrac{1}{2}\theta h^2 (\theta - 1) f''(x_0) + \cdots.$$

The first term in the series, namely $\tfrac{1}{2}\theta h^2 (\theta - 1) f''(x_0)$, is usually the most important one and is called the *principal truncation error*. When h is small (compared with one) the principal truncation error will normally give a satisfactory estimate of the total truncation error.

---

* See Appendix 1.

It must be emphasized, however, that evaluating $f''(x_0)$ is often very difficult and unsatisfactory (especially when the function itself is not known but only the function values).

EXERCISES

1. From the data given on page 40, determine approximations to log $83\cdot5$, log $81\cdot33$ and log $84\cdot25$ using linear interpolation.

2. Given that, correct to four significant figures, $\sqrt[3]{150} = 5\cdot313$ and $\sqrt[3]{160} = 5\cdot429$, obtain using linear interpolation an approximation to $\sqrt[3]{153\cdot8}$. (To four significant figures $\sqrt[3]{153\cdot8} = 5\cdot358$.)

3. Given that, to four significant figures, $e^{1\cdot75} = 5\cdot755$ and $e^{1\cdot80} = 6\cdot050$ obtain, by linear interpolation an estimate of $e^{1\cdot784}$. (To four significant figures $e^{1\cdot784} = 5\cdot954$.)

4. From the values of $1/x^2$ correct to 3 significant figures when $x = 2$ and $x = 3$ obtain an estimate of $1/(2\cdot54)^2$ using linear interpolation. (Correct to three significant figures $1/(2\cdot54)^2 = 0\cdot155$.)

5.2. Quadratic interpolation—
Newton's forward-difference formula of degree two

Instead of joining two neighbouring points with a straight line and using this for interpolation at intermediate points as in the previous section, we can join three neighbouring points with a quadratic and use this curve for interpolating. This method is known as *quadratic interpolation*.

To begin with we must be given at least three values of x, say x_0, x_1, x_2 and their corresponding function values $f(x_0), f(x_1), f(x_2)$. The first step is then to determine an equation for the quadratic curve passing through the three points $(x_0, f(x_0))$, $(x_1, f(x_1))$, $(x_2, f(x_2))$. This could be done simply by substituting in turn the coordinates of these points in an equation of the form $y = a_0 + a_1x + a_2x^2$ and solving the resulting simultaneous equations

$$a_0 + a_1x_0 + a_2x_0^2 = f(x_0)$$
$$a_0 + a_1x_1 + a_2x_1^2 = f(x_1)$$
$$a_0 + a_1x_2 + a_2x_2^2 = f(x_2)$$

for a_0, a_1 and a_2. However, the amount of work involved can be considerably reduced if we start with the quadratic equation in the form

$$y = A_0 + A_1(x - x_0) + A_2(x - x_0)(x - x_1).$$

(Note that the right-hand side of this equation is second degree in x and can be written in the form $a_0 + a_1x + a_2x^2$.)

Now since the curve must pass through the point $(x_0, f(x_0))$

$$f(x_0) = A_0.$$

Since the curve also passes through the point $(x_1, f(x_1))$

$$f(x_1) = A_0 + A_1(x_1 - x_0)$$
$$= f(x_0) + A_1 h$$

where $x_1 = x_0 + h$.

Hence
$$A_1 = \frac{f(x_1) - f(x_0)}{h} = \frac{1}{h}\Delta f_0.$$

Since the curve also passes through the point $(x_2, f(x_2))$

$$f(x_2) = A_0 + A_1(x_2 - x_0) + A_2(x_2 - x_0)(x_2 - x_1)$$
$$= f(x_0) + \frac{1}{h}\Delta f_0 (2h) + A_2(2h)(h)$$

where $x_2 = x_1 + h = x_0 + 2h$.

Hence
$$A_2 = \frac{1}{2h^2}\{f(x_2) - f(x_0) - 2\Delta f_0\}$$

$$= \frac{1}{2h^2}\{f(x_2) - 2f(x_1) + f(x_0)\}$$

$$= \frac{1}{2h^2}\Delta^2 f_0.$$

Therefore, we have

$$y = f_0 + \frac{x - x_0}{h}\Delta f_0 + \frac{(x - x_0)(x - x_1)}{2h^2}\Delta^2 f_0.$$

Now, when $x = x_0 + \theta h$ the corresponding value of y (that is the approximation y_θ to $f(x_0 + \theta h)$) is given by

$$y_\theta = f_0 + \theta \Delta f_0 + \tfrac{1}{2}\theta(\theta - 1)\Delta^2 f_0.$$

If possible x_0 should be chosen so that θ lies between 0 and 1. The approximate formula

$$f(x_0 + \theta h) \simeq f_0 + \theta \Delta f_0 + \tfrac{1}{2}\theta(\theta - 1)\Delta^2 f_0$$

is called Newton's forward-difference interpolation formula of degree two. Expressions for round-off errors and truncation errors can be obtained in the same way as in the previous sections.

Example. Use Newton's forward-difference interpolation formula of degree two and the values of $\cos 30°$, $\cos 60°$ and $\cos 90°$ to obtain an estimate of the value of $\cos 50°$.

$x°$	$\cos x$	$\Delta(\cos x)$	$\Delta^2(\cos x)$
30	0·866		
		-366	
60	0·500		-134
		-500	
90	0·000		

Here $h = 30°$ and $\theta = \frac{20}{30} = \frac{2}{3}$.

Hence $\cos 50° \simeq 0·866 + \frac{2}{3}(-0·366) + \frac{1}{2}\cdot\frac{2}{3}(-\frac{1}{3})(-0·134) \simeq 0·637$.

From tables the value of $\cos 50°$ is 0·643.

Note that, using linear interpolation, we obtain

$$\cos 50° \simeq 0·866 + \frac{2}{3}(-0·366) = 0·622.$$

Thus, in this case, quadratic interpolation has produced the better approximation.

EXERCISES

1. Given that, correct to four significant figures $\sqrt[3]{150} = 5·313$, $\sqrt[3]{160} = 5·429$ and $\sqrt[3]{170} = 5·540$, obtain using quadratic interpolation an approximation to $\sqrt[3]{153·8}$. Compare the approximation with that obtained in exercise 2 on page 46 and with the true value, correct to four significant figures, 5·358.

2. Given that, to four significant figures, $e^{1·70} = 5·474$, $e^{1·75} = 5·755$, $e^{1·80} = 6·050$ and $e^{1·85} = 6·360$, obtain using quadratic interpolation estimates of $e^{1·784}$ and $e^{1·725}$. Compare the estimate of $e^{1·784}$ obtained here with that obtained in exercise 3 on page 46 and with the true value, correct to four significant figures, 5·954.

3. From the values of $1/x^2$ correct to three decimal places, when $x = 2, 3, 4$, obtain an estimate of $1/(2·54)^2$ using quadratic interpolation. Compare this approximation with that obtained in exercise 4 on page 46 and with the true value, correct to three significant figures, 0·155.

5.3. Higher-order formulae

In a similar way we can join four neighbouring points with a cubic, five neighbouring points with a quartic, and so on, and then use these curves for interpolating at intermediate points; that is, we can develop cubic and quartic interpolation formulae and so on. Newton's forward-difference formula of degree n is

$$y_\theta = f_0 + \theta\Delta f_0 + \tfrac{1}{2}\theta(\theta-1)\Delta^2 f_0 + \ldots + \frac{1}{n!}\theta(\theta-1)\ldots(\theta-n+1)\Delta^n f_0.$$

It does *not* follow, however, that using a higher-order interpolating formula will necessarily give more accurate interpolated values. Whether it does or not depends upon the shape of the curve on which the given data should lie. One way of obtaining an indication of the degree of interpolating polynomial which should be used in a particular situation is by considering the differences of the given function values as in §4.4.

Example. Given the data

x	$f(x)$
1·0	2·98
1·2	3·55
1·4	4·45
1·6	5·63
1·8	7·16
2·0	9·01

use interpolation to obtain an approximation to the value of the function f at 1·5.

In order to obtain an indication of which degree of interpolating polynomial should be used, we first form a difference table for the given data.

x	f			
1·0	2·98			
		57		
1·2	3·55		33	
		90		
1·4	4·45		28	
		118		
1·6	5·63		35	
		153		
1·8	7·16		32	
		185		
2·0	9·01			

From this we see that the second differences are approximately constant. (The third differences are alternately positive and negative and the magnitudes of the higher-order differences increase.) Thus the given data are most closely approximated by a second degree polynomial and so a second degree interpolating polynomial should be used.

Newton's forward-difference interpolating polynomial of degree two is

$$f_0 + \theta \Delta f_0 + \tfrac{1}{2}\theta(\theta - 1)\Delta^2 f_0.$$

Hence, using this polynomial for interpolating with $f_0 = f(1\cdot4)$ and $\theta = \tfrac{1}{2}$ we obtain as our approximation for $f(1\cdot5)$,

$$4\cdot45 + \tfrac{1}{2}(1\cdot18) + \tfrac{1}{2}(\tfrac{1}{2})(-\tfrac{1}{2})(0\cdot35) \simeq 5\cdot00$$

(rounded to two decimal places).

Example. By differencing the following table, find the degree of the polynomial of minimum degree which will exactly fit the data

x	0	0·5	1·0	1·5	2·0	2·5	3·0
$f(x)$	0·25	−1·50	−1·75	−0·50	2·25	6·50	12·25

and obtain this polynomial using Newton's forward-difference formula of appropriate degree.

x	$f(x)$		
0	0·25		
		− 175	
0·5	−1·50		150
		−25	
1·0	−1·75		150
		125	
1·5	−0·50		150
		275	
2·0	2·25		150
		425	
2·5	6·50		150
		575	
3·0	12·25		

Since the second differences are constant, the required polynomial must have degree two. This polynomial can now be determined by obtaining Newton's forward-difference interpolating polynomial of degree two for the interval starting at 0. The general expression is

$$f_0 + \theta\Delta f_0 + \tfrac{1}{2}\theta(\theta-1)\Delta^2 f_0$$

from which we obtain here

$$0{\cdot}25 + \frac{x}{0{\cdot}5}(-1{\cdot}75) + \tfrac{1}{2}\frac{x}{0{\cdot}5}\left(\frac{x}{0{\cdot}5}-1\right)(1{\cdot}50)$$

since $\theta = x/h$ and $h = 0{\cdot}5$. On simplification this gives the polynomial

$$3x^2 - 5x + 0{\cdot}25.$$

Note that because the polynomial exactly fits all the data the same result is obtained using Newton's formula for any of the given intervals (provided of course the appropriate first and second differences are known). Care must be taken in the interpretation of θ however. For example, for the interval starting at 1·5,

$$\theta = \frac{x-1{\cdot}5}{0{\cdot}5}$$

and so we obtain the polynomial

$$-0.50 + \frac{x-1.5}{0.5}(2.75) + \tfrac{1}{2}\left(\frac{x-1.5}{0.5}\right)\left(\frac{x-1.5}{0.5}-1\right)(1.50).$$

On simplification this again reduces to the polynomial $3x^2 - 5x + 0.25$. This example should be compared with that in §4.4 on page 36.

EXERCISES

1. Given the data

x	1·0	1·1	1·2	1·3	1·4	1·5
$f(x)$	8·01	9·69	11·56	13·61	15·84	18·26

use an interpolation formula of appropriate degree to obtain an approximation to the value of the function f at 1·36.

2. Given the data

x	0	·2	·4	·6	·8	1·0
$f(x)$	0·782	1·033	1·283	1·534	1·784	2·035

use an interpolation formula of appropriate degree to obtain an approximation to the value of the function f at 0·452.

3. By differencing the following table, find the degree of the polynomial of minimum degree which will exactly fit the data

x	0	·5	1·0	1·5	2·0	2·5	3·0
$f(x)$	0·75	−0·25	−0·25	0·75	2·75	5·75	9·75

and obtain this polynomial using Newton's forward-difference formula of appropriate degree.

4. By differencing the following table, find the degree of the polynomial of minimum degree which will exactly fit the data

x	1·0	1·2	1·4	1·6	1·8	2·0
$f(x)$	2·25	1·73	0·97	−0·03	−1·27	−2·75

and obtain this polynomial using Newton's forward-difference formula of appropriate degree.

Integration

IN THIS CHAPTER WE WILL BE CONCERNED WITH THE DETERMINATION OF approximate numerical methods of integration. These methods are especially useful in the following three situations:

1. The function to be integrated is such that no analytical method exists. For example $\int_a^b \sqrt{\sin x}\, dx$.

2. An analytical method exists but is rather complex. For example

$$\int_a^b \frac{1}{1+x^4}\, dx$$

(see Chapter 1, page 2).

3. The function to be integrated is not known explicitly but instead we are given only a series of function values for values of the independent variable in the interval of integration $[a, b]$.

For cases 1 and 2 the first step is to evaluate the function f, which is to be integrated, for a series of values of the independent variable covering the interval of integration $[a, b]$; that is, cases 1 and 2 are first reduced to case 3.

Now the series of values of the independent variable x and the corresponding values of the function f can be plotted on graph paper as in Figure 3.

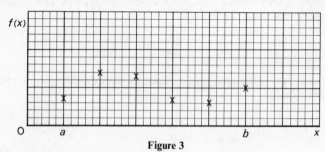

Figure 3

Then, if the shape of the curve $y = f(x)$ is known, it can be drawn (Figure 4) and the area under this curve between $x = a$ and $x = b$ (that is, the integral of $f(x)$ with respect to x from $x = a$ to $x = b$) estimated.

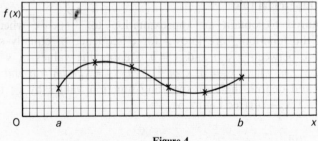

Figure 4

One way of estimating this area is to count the number of squares of the graph paper which are included in the area. An allowance has of course to be made for all the incomplete squares included (Figure 4). An upper bound to the integral is obtained by counting each incomplete square as if it were a complete square, and a lower bound is obtained by simply neglecting any incomplete squares.

Very often (as in Chapter 5) we know nothing about the shape of the curve represented by $y = f(x)$ in the interval from $x = a$ to $x = b$, and so we start by making some assumption as to its shape. The easiest assumption to make is that the curve can be approximated by a straight line between any two consecutive points (Figure 5a). Alternatively we may assume that the points, taken in threes, lie approximately on a parabola-type curve (that is, a curve with an equation of the second degree) which may be different for each set of three points as in Figure 5b and so on.

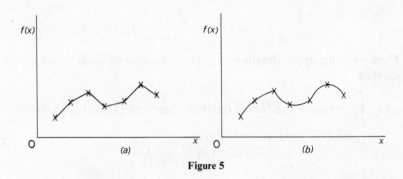

Figure 5

The different approximations we make as to the shape of the curve will lead to different numerical integration formulae and hence to various approximations to the integral.

6.1. The trapezoidal rule

It is required to obtain an approximation to $\int_a^b f(x)\,dx$ where the values of the function f are known for a sequence of equally spaced values of x between $x = a$ and $x = b$. Denote the values of x by x_r $(r = 0, 1, 2, \ldots, n)$ where $x_0 = a$, $x_r = x_0 + rh$, $x_n = x_0 + nh = b$ and h is a constant, and denote the corresponding function values by f_r; that is $f_r \equiv f(x_r) \equiv f(x_0 + rh)$ (see Figure 6).

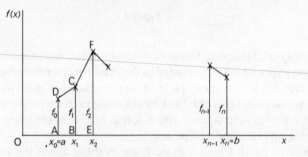

Figure 6

Since we do not know the shape of the graph of $y = f(x)$ we shall use as a first approximation the curve obtained by joining each consecutive pair of points (x_r, f_r) and (x_{r+1}, f_{r+1}) for $r = 0, 1, 2, \ldots, (n-1)$ by a straight line (that is, a curve of degree one) as in Figure 6.

Now the equation of the straight line joining the points (x_0, f_0) and (x_1, f_1) is

$$y = f_0 + (x - x_0)\left(\frac{f_1 - f_0}{x_1 - x_0}\right).$$

Then with this approximation to $f(x)$ in the interval from x_0 to x_1, we see that

$$\int_{x_0}^{x_1} f(x)\,dx \simeq \text{the area of the trapezium ABCD (see Figure 6)}$$

$$= \int_{x_0}^{x_1} \left\{ f_0 + (x - x_0)\left(\frac{f_1 - f_0}{x_1 - x_0}\right) \right\} dx$$

$$= f_0(x_1 - x_0) + \tfrac{1}{2}(x_1 - x_0)^2 \left(\frac{f_1 - f_0}{x_1 - x_0}\right)$$

$$= \tfrac{1}{2}h(f_0 + f_1).$$

Similarly

$$\int_{x_1}^{x_2} f(x)dx \simeq \text{the area of the trapezium BEFC (see Figure 6)}$$

$$= \tfrac{1}{2}h(f_1 + f_2).$$

Hence, summing over all the trapezia between $x = a$ and $x = b$, we see that

$$\int_a^b f(x)dx \equiv \int_{x_0}^{x_n} f(x)dx$$

$$= \int_{x_0}^{x_1} f(x)dx + \int_{x_1}^{x_2} f(x)dx + \cdots + \int_{x_{n-1}}^{x_n} f(x)dx$$

$$\simeq \tfrac{1}{2}h(f_0 + f_1) + \tfrac{1}{2}h(f_1 + f_2) + \cdots + \tfrac{1}{2}h(f_{n-1} + f_n)$$

$$= \tfrac{1}{2}h(f_0 + 2f_1 + 2f_2 + \cdots + 2f_{n-1} + f_n).$$

This is the *trapezoidal rule* and is of course exact if the curve represented by $y = f(x)$ is a straight line or a series of straight lines as in Figure 6.

Example. Use the trapezoidal rule to obtain from the data contained in the following table an approximation to $\int_2^4 f(x)dx$.

x	$f(x)$
2·0	1·7321
2·5	1·8708
3·0	2·0000
3·5	2·1213
4·0	2·2361

Here $h = 0.5$ and so from the trapezoidal rule we obtain

$$\int_2^4 f(x)dx \simeq \tfrac{1}{2} \times 0.5(1.7321 + 2 \times 1.8708 + 2 \times 2.0000 + 2 \times 2.1213 + 2.2361)$$

$$= 0.25(15.9524)$$

$$= 3.9881.$$

This result is of very limited value as it stands, since we do not know how good an approximation it is to the true value of the integral. How large is the error in the result due to the round-off errors in the original data? How large is the truncation error in this case?

We must now investigate the errors involved in applying the trapezoidal rule. Once again we shall not consider "mistakes" but shall only look at

C

the error due to round-off (or other) errors in the given data and the truncation error of the method (i.e. the error due to approximating the curve by the straight lines).

6.1.1. Effect of round-off on the trapezoidal rule

Let the error in f_r be e_r for $r = 0, 1, 2, \ldots, n$. Then the modulus of the error in the integral due to these errors in the given data is less than or equal to $\frac{1}{2}h(|e_0| + 2|e_1| + 2|e_2| + \cdots + 2|e_{n-1}| + |e_n|)$. Now if each of the absolute errors $|e_r|$, $r = 0, 1, 2, \ldots, n$, is less than or equal to $|e|$, as for example when they are entirely due to rounding the given data to a fixed number of decimal places, the modulus of the error in the integral is less than or equal to

$$\frac{1}{2}h(|e| + 2|e| + \cdots + 2|e| + |e|) = nh|e|$$
$$= (b-a)|e|.$$

Hence, if the errors in the function values f_r, $r = 0, 1, 2, \ldots, n$ are entirely due to rounding-off to k decimal places, then $|e| \leqslant \frac{1}{2}10^{-k}$ and so the modulus of the corresponding error in the integral is less than or equal to $\frac{1}{2}(b-a)10^{-k}$. Note that this bound on the error is independent of n and h.

Example. Estimate the error in the result of the last example due to possible round-off errors in the given function values.

Since the function values are given to four decimal places, the maximum absolute error in each is $\frac{1}{2}10^{-4}$.

Hence the modulus of the error in the integral value, due to this source, is less than or equal to $(4-2)\frac{1}{2}10^{-4} = 10^{-4} = 0.0001$.

Thus, even if there were no truncation error, the last digit in the value 3·9881 could not be regarded as accurate.

6.1.2. Truncation error in the trapezoidal rule

As for interpolation, the truncation error is more difficult to estimate than the round-off error. However, if the function f is sufficiently differentiable it can be shown, using Taylor series, that the modulus of the truncation error for the integral of f from a to b is less than or equal to

$$\tfrac{1}{12}h^2(b-a)|f''(\xi)|$$

where $$|f''(\xi)| = \max\{|f''(x)|; a \leqslant x \leqslant b\}.$$

Further, if it is known that f'' is always positive in the interval from a to b, then the truncation error will be negative, while if f'' is always negative then the truncation error will be positive.

Clearly, the smaller the value of h, then the smaller will be the modulus of the maximum truncation error.

Example. Given that the function $f(x)$ whose values are given in the example on page 55 is $(1+x)^{1/2}$, estimate the truncation error in the value of the integral obtained in that example.

Denote the truncation error by E. Then in the notation used above,

$$|E| \leqslant \tfrac{1}{12}h^2(b-a)|f''(\xi)|$$
$$= \tfrac{1}{12}h^3 n|f''(\xi)|.$$

Here $f(x) = (1+x)^{1/2}.$

$$f'(x) = \tfrac{1}{2}(1+x)^{-1/2} \quad \text{and} \quad f''(x) = -\tfrac{1}{4}(1+x)^{-3/2} = \frac{-1}{4\sqrt{(1+x)^3}}.$$

Clearly then $f''(x)$ has its maximum numerical value in the interval $2 \leqslant x \leqslant 4$ when $x = 2$; that is $\xi = 2$ and so

$$|f''(\xi)| = \frac{1}{4\sqrt{3^3}} = \frac{1}{12\sqrt{3}}.$$

Hence $$|E| \leqslant \tfrac{1}{12}(0{\cdot}5)^3\, 4\left(\frac{1}{12\sqrt{3}}\right) \simeq 0{\cdot}002.$$

But, from the form of $f''(x)$, it is obvious that $f''(x)$ is always negative for $2 \leqslant x \leqslant 4$ so that the truncation error E will be positive.

Thus $0 \leqslant E \leqslant 0{\cdot}002.$

Comparing this with the round-off error for the same problem (obtained in the previous example) we see that in this case the truncation error is dominant (and so the round-off error may be neglected).

Hence we see that any digits after the third decimal place may be meaningless in the value obtained for the integral.

Therefore $3{\cdot}988 \leqslant \displaystyle\int_2^4 (1+x)^{1/2}dx \leqslant 3{\cdot}988 + 0{\cdot}002 = 3{\cdot}990.$

Therefore $\displaystyle\int_2^4 (1+x)^{1/2}dx = 3{\cdot}99$ to two decimal places.

From ordinary (non-numerical) integration, we have

$$\int_2^4 (1+x)^{1/2} dx = \left[\tfrac{2}{3}(1+x)^{3/2}\right]_2^4$$
$$= \tfrac{2}{3}(\sqrt{125} - \sqrt{27})$$
$$= 3\cdot989$$

correct to three decimal places.

Now it must be emphasized that it will not always be possible to carry out an analysis similar to that of this last example. In very many cases it may be no easy task to determine $f''(x)$ and then, even when this can be done, there still remains the problem of determining ξ, that is, the value of x in the interval of integration for which $f''(x)$ has its maximum value. The procedure which is normally adopted when the truncation error cannot be readily estimated is illustrated in the following sub-section.

It should also be pointed out that the estimates we have obtained for the errors are all maximum values, and the actual errors occurring in any particular case may in fact be much less than these estimated values. Thus, while the expression obtained above for the maximum truncation error is of considerable theoretical importance, its practical value is very limited indeed.

6.1.3. *Application of the trapezoidal rule*

We shall consider separately the cases when

1. the modulus of the maximum truncation error, $\tfrac{1}{12}h^3 n |f''(\xi)|$, can be readily evaluated and

2. the expression $\tfrac{1}{12}h^3 n |f''(\xi)|$ cannot be readily evaluated.

1. $\tfrac{1}{12}h^3 n |f''(\xi)|$ can be readily evaluated.

If the accuracy required in the value of the integral is known, then we can calculate a suitable value of h and the number of decimal places which should be retained in the function values in order to achieve this. This means that the work of actually evaluating the integral is then minimized. Using an unnecessarily small value of h would involve evaluating the function at unnecessarily many points, and retaining too many digits in the function values themselves would involve extra (and unnecessary) work in the calculations. This latter point is unimportant if the calculations are being carried out on a computer.

Example. Determine a suitable value of h and the number of decimal figures which should be retained in the values of $(1+x)^{1/2}$ in order to evaluate $\int_2^4 (1+x)^{1/2} dx$ correct to three decimal places.

If the values of $(1+x)^{1/2}$ are rounded to k decimal places, then the resulting absolute round-off error in the integral will be less than or equal to $(4-2)\frac{1}{2}10^{-k} = 10^{-k}$.

We must try to choose k so that this error will not affect the value of the integral when rounded to three decimal places. This should normally be the case if 10^{-k} does not exceed 0·00005. Hence use $k = 5$, that is, round-off the values of $(1+x)^{1/2}$ to five decimal places. The modulus of the error will now be less than or equal to 0·00001.

The modulus of the maximum truncation error is less than or equal to

$$\tfrac{1}{12}h^2(b-a)|f''(\xi)| = \tfrac{1}{12}h^22\left(\frac{1}{12\sqrt{3}}\right)$$

(as in the previous example). Once again we require this error not to exceed say 0·00005, that is

$$\frac{h^2}{72\sqrt{3}} \leqslant 0·00005$$

therefore $\qquad\qquad h^2 \leqslant (0·00005)72\sqrt{3} \simeq 0·0062$

therefore $\qquad\qquad h \leqslant 0·078.$

Hence, a convenient value of h would be 0·05.

(Note that 0·07, 0·06 would not be suitable since the length of the interval of integration (namely 2) is not exactly divisible by either of these numbers.)

With $h = 0·05$, the modulus of the truncation error is less than or equal to

$$\frac{(0·05)^2}{72\sqrt{3}} \simeq 0·00002.$$

If the integral is now evaluated with $h = 0·05$ and retaining five decimal places in the values of $(1+x)^{1/2}$, the error in the result should lie between $-0·00001$ and $0·00003$, which should enable the final value to be obtained correct to three decimal places.

Now if our function values are specified to a given accuracy, this will limit the accuracy we can obtain in the final result; in theory we could then calculate a critical value of h such that using values smaller than this critical value will not increase the accuracy of the result. The error due to round-off would be dominant.

Alternatively, if h is given, this will limit the accuracy of the final result, and in theory we could calculate the maximum number of decimal places it would be useful to keep in the function values. Using any more decimal places would not increase the accuracy of the final result. The truncation error would be dominant.

2. $\frac{1}{12}h^3 n |f''(\xi)|$ cannot be readily evaluated.

This is the case most likely to occur in practice. Indeed, even when $\frac{1}{12}h^3 n |f''(\xi)|$ can be evaluated, it will still often be preferable to proceed in the following way rather than as described above.

6.1.4. Interval halving method

From the form of the truncation error we see that it can be reduced by using a smaller value of h. The method to be adopted now is as follows.

Starting with a relatively large value of h evaluate an approximation to the integral. Now replace h by $\frac{1}{2}h$ and evaluate another approximation to the integral. Replace $\frac{1}{2}h$ by $\frac{1}{4}h$ and evaluate another approximation to the integral, and so on. If the result is required to k significant figures, then this process of successively halving the interval h in x should be continued until the $(k+1)$th significant figure is constant, or at least until the variations in this figure are so small that the correct rounding-off to k figures can be carried out. If the $(k+1)$th figure is five, however, it may be necessary to carry out a further step in the process and to look at the $(k+2)$th figure to determine whether the five in the $(k+1)$th place is itself a result of rounding up so that the correct rounding to k figures can be carried out. The need for considering at least the $(k+1)$th figure can be illustrated as follows:

Suppose $\int_a^b f(x)dx$ is required to two decimal places; using interval size h, the value 3·719 is obtained for the integral, and using interval size $\frac{1}{2}h$ the value 3·724 is obtained. Correct to two decimal places these two values are the same, namely 3·72, and it might be thought that the result has now been obtained correct to two decimal places. However, using an interval size $\frac{1}{4}h$ the result could quite easily be 3·727, which correct to two decimal places is 3·73.

Readers familiar with elementary computer programming will readily realize that this method of successively halving the interval h is especially suitable for carrying out on a computer and will require quite a short program.

While this method does not involve estimating the truncation error, it may still be necessary to estimate, as above, the error due to round-off errors in the given data to ensure that we are not demanding a greater accuracy in the final result for the integral than the given data (that is, the function values) are capable of supplying.

Example. Use the trapezoidal rule and values of $(1+x)^{1/2}$ correct to two decimal places to obtain an approximation to the integral $\int_2^4 (1+x)^{1/2}dx$.

The modulus of the error in the integral due to round-off errors in the values of $(1+x)^{1/2}$ is less than or equal to $(4-2)\frac{1}{2}10^{-2} = 0.01$. Hence in the final result any digits after the second decimal place may be meaningless.

With $h = 2$ (see Figure 7a), we obtain the following values:

x	$(1+x)^{1/2}$
2	1·73
4	2·24

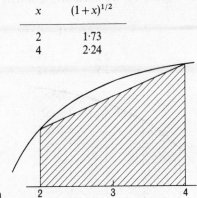

Figure 7a

Then
$$\int_2^4 (1+x)^{1/2}dx \simeq \tfrac{1}{2}(2)(1·73 + 2·24)$$

$$= 3·97.$$

Then, with $h = 1$ (see Figure 7b), we must evaluate $4^{1/2}$.

x	$(1+x)^{1/2}$
3	2·00

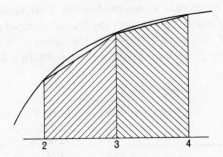

Figure 7b

Then $\qquad \displaystyle\int_2^4 (1+x)^{1/2}dx \simeq \frac{1}{2}(1)(1\cdot73+2(2\cdot00)+2\cdot24)$

$$= \tfrac{1}{2}(3\cdot97+2(2\cdot00))$$

$$= \tfrac{1}{2}(7\cdot97)$$

$$= 3\cdot985.$$

With $h=\frac{1}{2}$ (see Figure 7c) we must also evaluate $(3\cdot5)^{1/2}$ and $(4\cdot5)^{1/2}$.

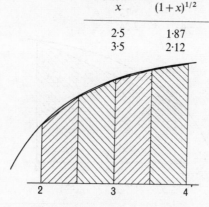

x	$(1+x)^{1/2}$
2·5	1·87
3·5	2·12

Figure 7c

Then $\qquad \displaystyle\int_2^4 (1+x)^{1/2}dx \simeq \frac{1}{2}(\tfrac{1}{2})(1\cdot73+2(1\cdot87)+2(2\cdot00)+2(2\cdot12)+2\cdot24)$

$$= \tfrac{1}{4}[7\cdot97+2(1\cdot87+2\cdot12)]$$

$$= \tfrac{1}{4}(15\cdot95)$$

$$= 3\cdot988.$$

This last number has been rounded to three decimal places since this is one digit more than is required in the final result. (We could have no confidence in any figures after the second decimal place in the final result because of the round-off errors in the values of $(1+x)^{1/2}$.)

With $h=\frac{1}{4}$ (see Figure 7d), we must obtain a few more values of the integrand

x	$(1+x)^{1/2}$
2·25	1·80
2·75	1·94
3·25	2·06
3·75	2·18

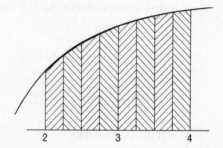

Figure 7d

$$\int_2^4 (1+x)^{1/2}dx \simeq \tfrac{1}{2}(\tfrac{1}{4})(1\cdot73 + 2(1\cdot80) + 2(1\cdot87) + 2(1\cdot94) +$$
$$2(2\cdot00) + 2(2\cdot06) + 2(2\cdot12) + 2(2\cdot18) + 2\cdot24)$$
$$= \tfrac{1}{8}[15\cdot95 + 2(1\cdot80 + 1\cdot94 + 2\cdot06 + 2\cdot18)]$$
$$= \tfrac{1}{8}(31\cdot91)$$
$$= 3\cdot989 \text{ rounded to three decimal places.}$$

At this stage it seems safe to assume that further applications of the process with $h = \tfrac{1}{8}, \tfrac{1}{16}, \ldots$ will all yield the value $3\cdot99$ rounded to two decimal places.

Hence we now have

$$\int_2^4 (1+x)^{1/2}dx = 3\cdot99 \pm 0\cdot01.$$

(Note that, correct to three decimal places, the true value of the integral is $3\cdot989$.)

Example. Evaluate

$$\int_{1\cdot5}^3 \frac{dx}{x-1}$$

correct to one decimal place using the trapezoidal rule.

If the values of $1/(x-1)$ are rounded to k decimal places, the modulus of the resultant error in the value of the integral is less than or equal to $(3-1\cdot5)\tfrac{1}{2}10^{-k} = \tfrac{3}{4}10^{-k}$. We now attempt to choose k so that this error will not affect the value of the integral when rounded to one decimal place. This should be the case if $\tfrac{3}{4}10^{-k}$ does not exceed $0\cdot005 = \tfrac{1}{2}10^{-2}$. The smallest integer value of k to satisfy this condition is 3. Thus we should evaluate the necessary values of $1/(x-1)$ correct to three decimal places.

We shall obtain a first approximation to the value of the integral by taking $h = \frac{3}{2}$.

x	$x-1$	$1/(x-1)$
$1\frac{1}{2}$	$\frac{1}{2}$	2·000
3	2	0·500

Then
$$\int_{1\cdot5}^{3} \frac{dx}{x-1} \simeq \tfrac{1}{2}(\tfrac{3}{2})(2\cdot000 + 0\cdot500)$$

$$= \tfrac{3}{4}(2\cdot500)$$

$$= 1\cdot88$$

rounded to two decimal places.

Now take $h = \frac{3}{4}$. We must calculate an additional value of the integrand.

x	$x-1$	$1/(x-1)$
$2\frac{1}{4}$	$1\frac{1}{4}$	0·800

Then
$$\int_{1\cdot5}^{3} \frac{dx}{x-1} \simeq \tfrac{1}{2}(\tfrac{3}{4})(2\cdot000 + 2(0\cdot800) + 0\cdot500)$$

$$= \tfrac{3}{8}(4\cdot100)$$

$$= 1\cdot54$$

rounded to two decimal places.

Now take $h = \frac{3}{8}$. We must calculate a few more values of the integrand.

x	$x-1$	$1/(x-1)$
$1\frac{7}{8}$	$\frac{7}{8}$	1·143
$2\frac{5}{8}$	$1\frac{5}{8}$	0·615

Then
$$\int_{1\cdot5}^{3} \frac{dx}{x-1} \simeq \tfrac{1}{2}(\tfrac{3}{8})(2\cdot000 + 2(1\cdot143 + 0\cdot800 + 0\cdot615) + 0\cdot500)$$

$$= \tfrac{3}{16}(4\cdot100 + 2(1\cdot143 + 0\cdot615))$$

$$= \tfrac{3}{16}(7\cdot616)$$

$$= 1\cdot43$$

rounded to two decimal places.

Now take $h = \frac{3}{16}$. We must calculate a few more values of the integrand.

x	$x-1$	$1/(x-1)$
$1\frac{11}{16}$	$\frac{11}{16}$	1·455
$2\frac{1}{16}$	$1\frac{1}{16}$	0·941
$2\frac{7}{16}$	$1\frac{7}{16}$	0·696
$2\frac{13}{16}$	$1\frac{13}{16}$	0·552

Then $\displaystyle\int_{1\cdot5}^{3} \frac{dx}{x-1} \simeq \frac{1}{2}\left(\frac{3}{16}\right)(7\cdot616 + 2(1\cdot455 + 0\cdot941 + 0\cdot696 + 0\cdot552))$

$$= \tfrac{3}{32}(14\cdot904)$$

$$= 1\cdot40$$

rounded to two decimal places.

Now take $h = \frac{3}{32}$. Once again we must calculate a few more values of the integrand.

x	$x-1$	$1/(x-1)$
$1\frac{19}{32}$	$\frac{19}{32}$	1·684
$1\frac{25}{32}$	$\frac{25}{32}$	1·280
$1\frac{31}{32}$	$\frac{31}{32}$	1·032
$2\frac{5}{32}$	$1\frac{5}{32}$	0·865
$2\frac{11}{32}$	$1\frac{11}{32}$	0·744
$2\frac{17}{32}$	$1\frac{17}{32}$	0·653
$2\frac{23}{32}$	$1\frac{23}{32}$	0·582
$2\frac{29}{32}$	$1\frac{29}{32}$	0·525

Then $\displaystyle\int_{1\cdot5}^{3} \frac{dx}{x-1} \simeq \frac{1}{2}\left(\frac{3}{32}\right)(14\cdot904 + 2(1\cdot684 + 1\cdot280 + 1\cdot032 + 0\cdot865 + 0\cdot744$

$$+ 0\cdot653 + 0\cdot582 + 0\cdot525))$$

$$= \tfrac{3}{64}(29\cdot634)$$

$$= 1\cdot39$$

rounded to two decimal places.

At this stage it seems safe to assume that further applications of the process with $h = \frac{3}{64}, \frac{3}{128}, \ldots$ will all yield the value 1·4 rounded to one decimal place.

Hence, correct to one decimal place

$$\int_{1\cdot5}^{3} \frac{dx}{x-1} = 1\cdot4.$$

(The value of the integral correct to three decimal places is 1·386.)

6.1.5. *Flow diagram*

In this subsection we give a simple flow diagram for applying the trapezoidal rule. Readers familiar with computer programming should be able to use this as a basis for writing a program.

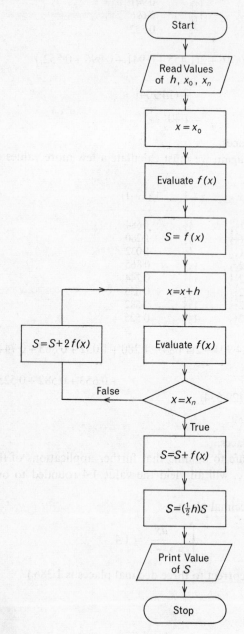

The reader should modify the above flow diagram so that it allows for a successive halving of the interval size h and subsequent re-evaluation of the integral until convergence is reached to some specified tolerance (as in the last example).

EXERCISES

1. From the following correctly rounded data calculate approximations to the value of the integral $\int_0^{0.8} f(x)dx$ using the trapezoidal rule with $h = 0.8, 0.4, 0.2, 0.1$.

x	0·0	0·1	0·2	0·3	0·4	0·5	0·6	0·7	0·8
$f(x)$	0·0000	0·0499	0·0995	0·1483	0·1960	0·2423	0·2867	0·3290	0·3688

2. Use the trapezoidal rule to obtain approximations to

$$\int_0^1 \frac{dx}{1+x}$$

taking (i) $h = 0.5$, (ii) $h = 0.25$ and (iii) $h = 0.125$ and working to five decimal places. Estimate round-off and truncation errors in each case.

3. Determine h so that the trapezoidal rule will give the value of

$$\int_3^{3.9} \frac{1}{x} dx$$

correct to three decimal places. How many decimal places should be retained in the values of $1/x$?

Evaluate the integral correct to four decimal places.

6.2. Simpson's Rule

Once again the problem is to determine an approximation to $\int_a^b f(x)dx$ where the values of the function f are known for a sequence of equally spaced values of x between $x = a$ and $x = b$ (Figure 8).

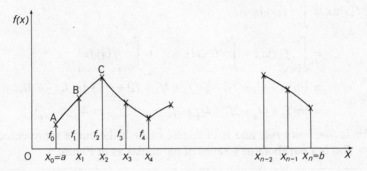

Figure 8

Instead of joining each consecutive pair of points (x_r, f_r), (x_{r+1}, f_{r+1}) by a straight line to obtain an approximation to the graph of $y = f(x)$ as in the derivation of the trapezoidal rule, we shall join the three points (x_0, f_0), (x_1, f_1) and (x_2, f_2) by a parabola (that is, a curve of degree two), the three points (x_2, f_2), (x_3, f_3) and (x_4, f_4) by another parabola, and so on up to the last three points (x_{n-2}, f_{n-2}), (x_{n-1}, f_{n-1}) and (x_n, f_n) (Figure 8). Note that this implies that n must be even, that is, that the number of points is odd. Note also that, using straight lines for the approximation, we could join each pair of consecutive points by a straight line, since it only requires two points to define a straight line uniquely. However, if we use parabolas (or curves of degree two) then it is necessary to take the points in threes since it requires three points to define a parabola uniquely.

As in §5.2 the parabola through the points $A(x_0, f_0)$, $B(x_1, f_1)$, $C(x_2, f_2)$ has the equation

$$y = f_0 + \frac{x - x_0}{h} \Delta f_0 + \frac{(x - x_0)(x - x_1)}{2h^2} \Delta^2 f_0.$$

Then, with this approximation to $f(x)$ in the interval from x_0 to x_2 we see that

$$\int_{x_0}^{x_2} f(x) dx \simeq \int_{x_0}^{x_2} \left\{ f_0 + \frac{x - x_0}{h} \Delta f_0 + \frac{(x - x_0)(x - x_1)}{2h^2} \Delta^2 f_0 \right\} dx$$

$$= \tfrac{1}{3} h (f_0 + 4f_1 + f_2).^*$$

Similarly
$$\int_{x_2}^{x_4} f(x) dx \simeq \tfrac{1}{3} h (f_2 + 4f_3 + f_4).$$

Hence

$$\int_a^b f(x) dx \equiv \int_{x_0}^{x_n} f(x) dx$$

$$= \int_{x_0}^{x_2} f(x) dx + \int_{x_2}^{x_4} f(x) dx + \cdots + \int_{x_{n-2}}^{x_n} f(x) dx$$

$$\simeq \tfrac{1}{3} h (f_0 + 4f_1 + f_2) + \tfrac{1}{3} h (f_2 + 4f_3 + f_4) + \cdots + \tfrac{1}{3} h (f_{n-2} + 4f_{n-1} + f_n)$$

$$= \tfrac{1}{3} h (f_0 + 4f_1 + 2f_2 + 4f_3 + 2f_4 + \cdots + 2f_{n-2} + 4f_{n-1} + f_n).$$

This is *Simpson's rule* and is of course exact if the curve represented by $y = f(x)$ is a parabola or a series of parabolas as in Figure 8.

---

* For details of the integration see Appendix 3.

Example. Use Simpson's rule to obtain from the data of the example on page 55 an approximation to $\int_2^4 f(x)dx$.

Here $h = 0.5$ and so, from Simpson's rule, we obtain

$$\int_2^4 f(x)dx \simeq \tfrac{1}{3} \times 0.5(1.7321 + 4(1.8708) + 2(2.000) + 4(2.1213) + 2.2361)$$

$$= \tfrac{1}{3} \times 0.5(23.9366)$$

$$= 3.9894 \text{ rounded to five significant digits.}$$

This result is of limited value until we can estimate its accuracy.

6.2.1. *Effect of round-off on Simpson's rule*

With the same notation as before, the error in the integral due to errors in the given data is equal to

$$\tfrac{1}{3}h(e_0 + 4e_1 + 2e_2 + \cdots + 4e_{n-1} + e_n).$$

Hence the modulus of this round-off error is less than or equal to

$$\tfrac{1}{3}h(|e_0| + 4|e_1| + 2|e_2| + \cdots + 4|e_{n-1}| + |e_n|).$$

Now if each of the absolute errors $|e_r|$, $r = 0, 1, 2, \ldots, n$, is less than or equal to $|e|$, then the modulus of the resulting error in the integral is less than or equal to

$$\tfrac{1}{3}h(|e| + 4|e| + 2|e| + \cdots + 4|e| + |e|) = \tfrac{1}{3}h(2 + 4(\tfrac{1}{2}n) + 2(\tfrac{1}{2}n - 1))|e|$$

$$= nh|e|$$

$$= (b-a)|e|.$$

Hence, if all the function values have been rounded-off to k decimal places, $|e| \leqslant \tfrac{1}{2}10^{-k}$ and the modulus of the corresponding error in the integral is less than or equal to $\tfrac{1}{2}(b-a)10^{-k}$. Note that this is the same expression as was obtained for the trapezoidal rule.

Example. Estimate the round-off error in the result of the last example due to possible round-off errors in the given function values.

The modulus of the error in the integral, due to the round-off errors in the given function values is less than or equal to $\tfrac{1}{2}(4-2)10^{-4} = 0.0001$.

6.2.2. *Truncation error in Simpson's rule*

If the function f is sufficiently differentiable it can be shown, using Taylor series, that the modulus of the truncation error for the integral of f from

a to b is less than or equal to

$$\tfrac{1}{180}h^4(b-a)|f^{iv}(\xi)|$$

where $$|f^{iv}(\xi)| = \max\{|f^{iv}(x)|; a \leqslant x \leqslant b\}.$$

Further, if it is known that f^{iv} is always positive in the interval from a to b, then the truncation error will be negative, while if f^{iv} is always negative then the truncation error will be positive.

Once again, the smaller the value of h, the smaller will be the modulus of the maximum truncation error.

Example. Estimate the truncation error in the value of $\int_2^4 (1+x)^{1/2}dx$ obtained in the example on page 69.

When $f(x) = (1+x)^{1/2}$, $f'(x) = \tfrac{1}{2}(1+x)^{-1/2}$, $f''(x) = -\tfrac{1}{4}(1+x)^{-3/2}$, $f'''(x) = \tfrac{3}{8}(1+x)^{-5/2}$ and $f^{iv}(x) = -\tfrac{15}{16}(1+x)^{-7/2}$.

Clearly then f^{iv} has its maximum numerical value in the interval $2 \leqslant x \leqslant 4$ when $x = 2$; that is, $\xi = 2$ and so

$$|f^{iv}(\xi)| = \tfrac{15}{16}\frac{1}{\sqrt{3^7}} = \frac{15}{432\sqrt{3}}.$$

Hence the modulus of the truncation error

$$\leqslant \tfrac{1}{180}h^4(b-a)|f^{iv}(\xi)|$$

$$= \tfrac{1}{180}(0 \cdot 5)^4(4-2)\frac{15}{432\sqrt{3}}$$

$$\simeq 0 \cdot 000016.$$

But, from the form of f^{iv} it is obvious that f^{iv} is always negative for $2 \leqslant x \leqslant 4$ so that the truncation error will be positive.

Thus, the truncation error will lie between 0 and $0 \cdot 00002$. Comparing this truncation error with the round-off error for the same problem (obtained in the previous example) we see that in this case the round-off error is dominant (and so the truncation error may be neglected).

Hence $$3 \cdot 9893 \leqslant \int_2^4 (1+x)^{1/2}dx \leqslant 3 \cdot 9895.$$

Therefore $$\int_2^4 (1+x)^{1/2}dx = 3 \cdot 989 \text{ to three decimal places.}$$

(The true value of the integral to five decimal places is $3 \cdot 98946$.)

It must be emphasized that it will very rarely be possible to carry out an analysis similar to that of this last example because of the difficulties in evaluating $f^{iv}(\xi)$.

6.2.3. *Application of Simpson's rule*

As for the trapezoidal rule, if the truncation error can be readily estimated, then we can determine a suitable value of h and the number of decimal places which should be retained in the function values in order to attain a given accuracy in the value of the integral. However, quite often it will be impractical to evaluate the truncation error. In such cases, we calculate successive approximations to the integral using the intervals $h, \frac{1}{2}h, \frac{1}{4}h, \ldots$ until the required accuracy is attained (as for the trapezoidal rule). This method will be illustrated in examples.

Example. From values of $(1+x)^{1/2}$ correct to two decimal places use Simpson's rule to obtain an approximation to the integral $\int_2^4 (1+x)^{1/2}\,dx$. (The values of $(1+x)^{1/2}$ used in this example are given in the example on page 61.)

The modulus of the error in the integral due to round-off errors in the values of $(1+x)^{1/2}$ is less than or equal to $(4-2)\frac{1}{2}10^{-2} = 0.01$. Hence in the final result any digits after the second decimal place may be meaningless.

Now, taking $h = 1$, we obtain

$$\int_2^4 (1+x)^{1/2}dx \simeq \tfrac{1}{3}(1.73 + 4(2.00) + 2.24)$$

$$= \tfrac{1}{3}(11.97) = 3.99.$$

Then, with $h = \frac{1}{2}$, we obtain

$$\int_2^4 (1+x)^{1/2}dx \simeq \tfrac{1}{3}(\tfrac{1}{2})(1.73 + 4(1.87) + 2(2.00) + 4(2.12) + 2.24)$$

$$= \tfrac{1}{6}(3.97 + 2 \times 2.00 + 4(1.87 + 2.12))$$

$$= \tfrac{1}{6}(23.93) = 3.988 \text{ rounded to three decimal places.}$$

With $h = \frac{1}{4}$, we obtain

$$\int_2^4 (1+x)^{1/2}dx \simeq \tfrac{1}{3}(\tfrac{1}{4})(1.73 + 4(1.80) + 2(1.87) + 4(1.94) + 2(2.00) + 4(2.06)$$

$$+ 2(2.12) + 4(2.18) + 2.24)$$

$$= \tfrac{1}{12}\{3.97 + 2(2.00 + 1.87 + 2.12)$$

$$+ 4(1.80 + 1.94 + 2.06 + 2.18)\}$$

$$= \tfrac{1}{12}(47.87) = 3.989 \text{ rounded to three decimal places.}$$

At this stage it seems safe to assume that further applications of the process with $h = \frac{1}{8}, \frac{1}{16}, \ldots$ will all yield the value 3·99 rounded to two decimal places.

Hence we now have

$$\int_2^4 (1+x)^{1/2} dx = 3·99 \pm 0·01.$$

Example. Evaluate

$$\int_{1·5}^3 \frac{dx}{x-1}$$

correct to one decimal place using Simpson's rule.

As in the example on page 63 we see that we should evaluate the necessary values of $1/(x-1)$ correct to three decimal places. (The evaluations obtained in that example will be used here.)

With $h = \frac{3}{4}$, we obtain

$$\int_{1·5}^3 \frac{1}{x-1} dx \simeq \tfrac{1}{3}(\tfrac{3}{4})(2·00 + 4(0·800) + 0·500)$$

$$= \tfrac{1}{4}(2·500 + 4(0·800))$$

$$= \tfrac{1}{4}(5·700) = 1·42 \text{ to two decimal places.}$$

With $h = \frac{3}{8}$, we obtain

$$\int_{1·5}^3 \frac{1}{x-1} dx \simeq \tfrac{1}{3}(\tfrac{3}{8})(2·500 + 2(0·800) + 4(1·143 + 0·615))$$

$$= \tfrac{1}{8}(11·132) = 1·39 \text{ to two decimal places.}$$

With $h = \frac{3}{16}$, we obtain

$$\int_{1·5}^3 \frac{1}{x-1} dx \simeq \tfrac{1}{3}(\tfrac{3}{16})(2·500 + 2(0·800 + 1·143 + 0·615)$$

$$+ 4(1·455 + 0·941 + 0·696 + 0·552))$$

$$= \tfrac{1}{16}(22·192) = 1·39 \text{ to two decimal places.}$$

Hence $\quad \int_{1·5}^3 \frac{1}{x-1} dx = 1·4 \text{ correct to one decimal place.}$

6.2.4. *Flow diagram*

The flow diagram for applying Simpson's rule is similar to that for applying the trapezoidal rule given in §6.1.5 and the reader should not have too much difficulty in carrying out the appropriate modifications.

Having obtained a flow diagram for Simpson's rule similar to that given in §6.1.5 for the trapezoidal rule, the reader should then modify it to allow for a successive halving of the interval size h and subsequent re-evaluation of the integral until convergence is reached to some specified tolerance.

EXERCISES

1. From the following correctly rounded data calculate approximations to the value of the integral $\int_0^4 f(x)dx$ using Simpson's rule with $h = 2, 1, \frac{1}{2}$ and compare the results with the rounded true value of $1\cdot3971$.

x	0·0	0·5	1·0	1·5	2·0	2·5	3·0	3·5	4·0
$f(x)$	0·0000	0·2423	0·4401	0·5579	0·5767	0·4971	0·3391	0·1374	−0·0660

2. Use Simpson's rule to obtain approximations to

$$\int_0^1 \frac{dx}{1+x}$$

taking (i) $h = 0\cdot5$, (ii) $h = 0\cdot25$ and (iii) $h = 0\cdot125$ and working to five decimal places.

Estimate round-off and truncation errors in each case. Compare the results with those obtained in Exercise 2 on page 67.

3. Determine h so that Simpson's rule will give the value of

$$\int_3^{3\cdot9} \frac{1}{x} dx \text{ correct to three decimal places.}$$

How many decimal places should be retained in the values of $1/x$?
Evaluate the above integral correct to three decimal places.

6.3. Comparison of the trapezoidal rule and Simpson's rule

We have seen above that the effect of round-off errors on the two methods is similar. Using the trapezoidal rule the truncation error is proportional to h^2, while using Simpson's rule the truncation error is proportional to h^4. But for values of h which are less than 1, h^4 is smaller than h^2 and so in general, unless f'' is very small or f^{iv} is very large, we expect Simpson's rule to be more accurate than the trapezoidal rule for a given value of $h < 1$.

Example. With $h = \frac{1}{2}$ estimate

$$\int_0^1 \frac{1}{1+x^2} dx$$

using (i) the trapezoidal rule and (ii) Simpson's rule.

Compare the results of (i) and (ii) with the value 0·785 which is the true value of the integral rounded to three decimal places.

$$\text{If } f(x) = \frac{1}{1+x^2} \quad \text{then}$$

$$f(0) = 1$$
$$f(\tfrac{1}{2}) = 0·8$$
$$f(1) = 0·5.$$

Note that all these function values are exact (and not subject to round-off errors).

(i) $$\int_0^1 \frac{1}{1+x^2} dx \simeq \tfrac{1}{2}(\tfrac{1}{2})(1+2(0·8)+0·5)$$

$$= \tfrac{1}{4}(3·1) = 0·775.$$

Note that this result is free from any round-off error.

(ii) $$\int_0^1 \frac{1}{1+x^2} dx \simeq \tfrac{1}{3}(\tfrac{1}{2})(1+4(0·8)+0·5)$$

$$= \tfrac{1}{6}(4·7) = 0·783 \text{ rounded to three decimal places.}$$

Comparing both of these approximations with the true value (to three decimal places) of 0·785 it is clear that Simpson's rule has given the better approximation.

EXERCISES

1. From the given correctly rounded data evaluate the best approximation you can to $\int_0^{1·6} e^x \, dx$ using (i) the trapezoidal rule and (ii) Simpson's rule.

x	0	0·2	0·4	0·6	0·8	1·0	1·2	1·4	1·6
e^x	1	1·2214	1·4918	1·8221	2·2255	2·7183	3·3201	4·0552	4·9530

How many figures would you trust in your answers?

2. Evaluate $\int_0^{0·8} \tan x \, dx$ correct to three decimal places using (i) the trapezoidal rule and (ii) Simpson's rule.

3. Evaluate

$$\int_0^2 \frac{x+1}{x^2+1} dx$$

correct to two decimal places using (i) the trapezoidal rule and (ii) Simpson's rule.

6.4. The method of undetermined coefficients

In this section we consider an important and widely applicable alternative approach to the derivation of integration formulae.

6.4.1. Newton-Cotes-type formulae

We shall look here for integration formulae which express the approximation to the definite integral $\int_0^h f(x)\,dx$ in the general form

$$\alpha f(0) + \beta f(h) + \gamma f(2h) + \cdots + E$$

where α, β, γ, ... are undetermined coefficients and E is an error term. This is a perfectly reasonable starting point since all we are setting out to do is obtain an approximation to the integral as a linear combination of the values of the integrand. (Note that we have already seen that both the trapezoidal rule and Simpson's rule are of this form.) Formulae of the above form are called Newton-Cotes formulae.

In particular, considering the formula

$$\int_0^h f(x)\,dx = \alpha f(0) + E$$

we see that the error term can be made zero for the function $f(x) = 1$ (and so for all constant functions) if α is chosen appropriately as follows:

$$\int_0^h f(x)\,dx = \alpha f(0) + E.$$

Now put $f(x) = 1$.

Therefore

$$\int_0^h 1\,dx = \alpha + E.$$

But

$$\int_0^h 1\,dx = h \text{ and so } E \text{ will be zero if } \alpha = h.$$

Thus we have the approximate integration formula

$$\int_0^h f(x)\,dx \simeq hf(0).$$

which is in fact exact if $f(x) = $ constant (Figure 9). When $f(x) \neq$ constant the error term $E(\neq 0)$ can be found using Taylor series.

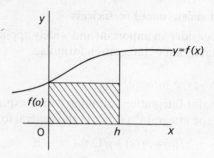

Figure 9

6.4.1.1. *The trapezoidal rule.* Consider the formula

$$\int_0^h f(x)dx = \alpha f(0) + \beta f(h) + E.$$

We now have two undetermined coefficients, α and β. Thus the error term E in this formula can be made zero for $f(x) = 1$ and for $f(x) = x$ by choosing α and β appropriately as follows:

$$\int_0^h 1 dx = h.$$

Therefore E in the above formula will be zero when $f(x) = 1$ if

$$\alpha + \beta = h.$$

Also
$$\int_0^h x dx = \left[\tfrac{1}{2}x^2\right]_0^h = \tfrac{1}{2}h^2.$$

Therefore E in the above formula will be zero when $f(x) = x$ if

$$\alpha \times 0 + \beta h = \tfrac{1}{2}h^2$$

since $f(0) = 0$ and $f(h) = h$.
Therefore $\beta = \tfrac{1}{2}h$ and $\alpha = \tfrac{1}{2}h$.
Hence we have the approximate integration formula

$$\int_0^h f(x)dx \simeq \tfrac{1}{2}hf(0) + \tfrac{1}{2}hf(h) = \tfrac{1}{2}h(f(0) + f(h)).$$

If $f(x)$ is *any* linear function, say $f(x) = lx + m$ where l and m are constants, then

$$\int_0^h f(x)\,dx = \int_0^h (lx + m)\,dx$$

$$= \left[\tfrac{1}{2}lx^2 + mx\right]_0^h$$

$$= \tfrac{1}{2}lh^2 + mh$$

$$= \tfrac{1}{2}h(lh + 2m)$$

$$= \tfrac{1}{2}h(m + (lh + m)) = \tfrac{1}{2}h(f(0) + f(h)).$$

Hence the formula is exact for all linear functions.

When $f(x)$ is not linear, the truncation error term $E(\neq 0)$ can be found using Taylor series.

Now put $u = x - x_0$ in the integral

$$\int_{x_0}^{x_1} f(x)\,dx$$

in which $x_1 = x_0 + h$.

Therefore

$$\int_{x_0}^{x_1} f(x)\,dx = \int_0^h f(u + x_0)\,du$$

$$= \int_0^h F(u)\,du$$

where $F(u) = f(u + x_0)$.

But

$$\int_0^h F(u)\,du \simeq \tfrac{1}{2}h\left[F(0) + F(h)\right]$$

$$= \tfrac{1}{2}h\left[f(x_0) + f(h + x_0)\right]$$

$$= \tfrac{1}{2}h\left[f(x_0) + f(x_1)\right].$$

Hence

$$\int_{x_0}^{x_1} f(x)\,dx \simeq \tfrac{1}{2}h\left[f(x_0) + f(x_1)\right].$$

This is the trapezoidal rule.

Then

$$\int_a^b f(x) = \int_{x_0}^{x_n} f(x)\,dx$$

$$= \int_{x_0}^{x_1} f(x)\,dx + \int_{x_1}^{x_2} f(x)\,dx + \ldots + \int_{x_{n-1}}^{x_n} f(x)\,dx$$

$$\simeq \tfrac{1}{2}h\left[f(x_0) + f(x_1)\right] + \tfrac{1}{2}h\left[f(x_1) + f(x_2)\right] + \cdots + \tfrac{1}{2}h\left[f(x_{n-1}) + f(x_n)\right]$$

$$= \tfrac{1}{2}h(f_0 + 2f_1 + 2f_2 + \cdots + 2f_{n-1} + f_n).$$

Example. Show that the trapezoidal rule is not exact for a quadratic function.

Take $f(x) = kx^2 + lx + m$.
Then we know that the trapezoidal rule is exact for the expression $(lx + m)$ and so we need only consider the integral of $g(x) = kx^2$.

$$\int_0^h g(x)dx = \int_0^h kx^2 dx = k\left[\tfrac{1}{3}x^3\right]_0^h$$

$$= \tfrac{1}{3}kh^3.$$

But $\tfrac{1}{2}h[g(0) + g(h)] = \tfrac{1}{2}h(kh^2) = \tfrac{1}{2}kh^3.$

Therefore $\int_0^h g(x)dx \neq \tfrac{1}{2}[g(0) + g(h)]$

and so the trapezoidal rule is not exact for a quadratic function. (Note that in this case the truncation error E is $\tfrac{1}{3}kh^3 - \tfrac{1}{2}kh^3 = -\tfrac{1}{6}kh^3 = -\tfrac{1}{12}h^3 g''$, compare §6.1.2, page 56.)

We can of course derive other formulae of the form

$$\int_0^h f(x)dx = \alpha f(0) + \beta f(h) + \gamma f(2h) + \cdots + E$$

in the same way as above.

For example, choosing α, β and γ appropriately in

$$\int_0^h f(x)dx = \alpha f(0) + \beta f(h) + \gamma f(2h) + E$$

we can obtain a formula for which the error term E is zero for all polynomial functions $f(x)$ with degree $\leqslant 2$ (that is, for constants, linear functions and quadratic functions). Such formulae are not in common use, however, except in special circumstances, and so will not be considered further.

6.4.1.2. Simpson's rule. Now consider the formula

$$\int_0^{2h} f(x)dx = \alpha f(0) + \beta f(h) + \gamma f(2h) + E.$$

We now have three undetermined coefficients α, β and γ and we can use these to ensure that the error term E is zero for $f(x) = 1$, $f(x) = x$ and $f(x) = x^2$.

$$\int_0^{2h} 1\,dx = 2h.$$

Hence E will be zero in the above formula when $f(x) = 1$ if

$$\alpha + \beta + \gamma = 2h. \qquad (1)$$

$$\int_0^{2h} x\,dx = [\tfrac{1}{2}x^2]_0^{2h} = 2h^2.$$

Then E will be zero in the above formula when $f(x) = x$ if

$$\alpha \times 0 + \beta h + \gamma(2h) = 2h^2$$

that is, if $\qquad\qquad\qquad \beta + 2\gamma = 2h. \qquad (2)$

Also $\qquad\qquad \displaystyle\int_0^{2h} x^2\,dx = [\tfrac{1}{3}x^3]_0^{2h} = \tfrac{8}{3}h^3.$

Therefore E will be zero in the above formula when $f(x) = x^2$ if

$$\alpha \times 0 + \beta h^2 + \gamma(2h)^2 = \tfrac{8}{3}h^3$$

that is, if $\qquad\qquad\qquad \beta + 4\gamma = \tfrac{8}{3}h. \qquad (3)$

Now, solving the three simultaneous equations (1), (2) and (3) for α, β and γ, we obtain $\alpha = \tfrac{1}{3}h$, $\beta = \tfrac{4}{3}h$ and $\gamma = \tfrac{1}{3}h$.

Hence we have the approximate integration formula

$$\int_0^{2h} f(x)\,dx \simeq \tfrac{1}{3}hf(0) + \tfrac{4}{3}hf(h) + \tfrac{1}{3}hf(2h)$$

$$= \tfrac{1}{3}h(f(0) + 4f(h) + f(2h)).$$

If $f(x)$ is *any* quadratic function, say $f(x) = kx^2 + lx + m$ where k, l and m are constants, then

$$\int_0^{2h} f(x)\,dx = \int_0^{2h} (kx^2 + lx + m)\,dx$$

$$= [\tfrac{1}{3}kx^3 + \tfrac{1}{2}lx^2 + mx]_0^{2h}$$

$$= \tfrac{8}{3}kh^3 + 2lh^2 + 2mh$$

$$= \tfrac{1}{3}h(f(0) + 4f(h) + f(2h)).$$

Hence the formula is exact for all quadratic functions (and so for all linear functions or constant functions).

Now consider $f(x) = x^3$.

$$\int_0^{2h} x^3 dx = \left[\tfrac{1}{4}x^4\right]_0^{2h} = 4h^4$$

$$= \tfrac{1}{3}h(4h^3 + (2h)^3)$$

$$= \tfrac{1}{3}h(f(0) + 4f(h) + f(2h)).$$

Hence the formula is also exact for $f(x) = x^3$ and so (as can be shown) for all cubic functions. Thus we have an unexpected "bonus".

Now consider $f(x) = x^4$.

$$\int_0^{2h} x^4 dx = \left[\tfrac{1}{5}x^5\right]_0^{2h} = \tfrac{32}{5}h^5.$$

But
$$\tfrac{1}{3}h(f(0) + 4f(h) + f(2h)) = \tfrac{1}{3}h(4h^4 + 16h^4)$$

$$= \tfrac{20}{3}h^5.$$

Therefore
$$\int_0^{2h} x^4\, dx \neq \tfrac{1}{3}h(f(0) + 4f(h) + f(2h))$$

and so the formula is not exact for a quartic function. (Note that in this case the truncation error E is $\tfrac{32}{5}h^5 - \tfrac{20}{3}h^5 = -\tfrac{4}{15}h^5 = -\tfrac{1}{90}h^5 f^{iv}$; compare §6.2.2, page 69.)

Hence the formula is exact for all polynomials of degree less than or equal to three. For other functions the form of the truncation error E can be found using Taylor series.

Now put $u = x - x_0$ in the integral

$$\int_{x_0}^{x_2} f(x)dx$$

in which $x_2 = x_1 + h = x_0 + 2h$.

Therefore
$$\int_{x_0}^{x_1} f(x)dx = \int_0^{2h} f(u + x_0)\, du$$

$$= \int_0^{2h} F(u)du$$

where $F(u) = f(u + x_0)$.

But $\qquad \int_0^{2h} F(u)du \simeq \tfrac{1}{3}h[F(0)+4F(h)+F(2h)]$

$$= \tfrac{1}{3}h[f(x_0)+4f(x_0+h)+f(x_0+2h)]$$

$$= \tfrac{1}{3}h[f(x_0)+4f(x_1)+f(x_2)].$$

Hence $\qquad \int_{x_0}^{x_2} f(x)dx \simeq \tfrac{1}{3}h[f(x_0)+4f(x_1)+f(x_2)].$

This is Simpson's rule.

Then $\qquad \int_a^b f(x)dx = \int_{x_0}^{x_n} f(x)dx$

$$= \int_{x_0}^{x_2} f(x)dx + \int_{x_2}^{x_4} f(x)dx + \cdots + \int_{x_{n-2}}^{x_n} f(x)dx$$

$$\simeq \tfrac{1}{3}h[f(x_0)+4f(x_1)+f(x_2)] + \tfrac{1}{3}h[f(x_2)+4f(x_3)+f(x_4)]$$

$$+ \cdots + \tfrac{1}{3}h[f(x_{n-2})+4f(x_{n-1})+f(x_n)]$$

$$= \tfrac{1}{3}h(f_0+4f_1+2f_2+4f_3+\cdots+2f_{n-2}+4f_{n-1}+f_n).$$

We can of course derive other formulae of the form

$$\int_0^{2h} f(x)dx = \alpha f(0)+\beta f(h)+\gamma f(2h)+\cdots+E$$

in the same way as above. Such formulae are not in common use, however, except in special circumstances, and so will not be considered further.

6.4.2. Euler-Maclaurin formulae

The method of undetermined coefficients can also be applied to obtain approximate integration formulae in terms of the values of the integrand and its derivative at the equally spaced tabulation points. Such formulae (involving integrand derivative values as well as integrand values) are called Euler-Maclaurin formulae.

To illustrate the procedure we shall apply the method of undetermined coefficients to derive a formula of the type

$$\int_0^h f(x)dx = \alpha f(0)+\beta f(h)+\alpha_1 f'(0)+\beta_1 f'(h)+E$$

in which α, α_1, β, β_1 are constants and E is an error term.

We have four undetermined coefficients α, α_1, β, β_1 so that the error term E in the above formula can be made zero for $f(x) = 1$, $f(x) = x$, $f(x) = x^2$ and $f(x) = x^3$ by choosing these coefficients appropriately.

$$\int_0^h 1 \, dx = h.$$

Therefore E in the above integration formula will be zero when $f(x) = 1$ and $f'(x) = 0$ if

$$\alpha + \beta = h.$$

$$\int_0^h x \, dx = \tfrac{1}{2}h^2.$$

Therefore E in the above integration formula will be zero for $f(x) = x$ and $f'(x) = 1$ if

$$h\beta + \alpha_1 + \beta_1 = \tfrac{1}{2}h^2.$$

$$\int_0^h x^2 \, dx = \tfrac{1}{3}h^3.$$

Therefore E in the above integration formula will be zero for $f(x) = x^2$ and $f'(x) = 2x$ if

$$h^2\beta + 2h\beta_1 = \tfrac{1}{3}h^3.$$

$$\int_0^h x^3 \, dx = \tfrac{1}{4}h^4.$$

Therefore E in the above integration formula will be zero for $f(x) = x^3$ and $f'(x) = 3x^2$ if

$$h^3\beta + 3h^2\beta_1 = \tfrac{1}{4}h^4.$$

Then we obtain, on solving the simultaneous equations for α, α_1, β, β_1,

$$\alpha = \tfrac{1}{2}h, \quad \beta = \tfrac{1}{2}h, \quad \alpha_1 = \tfrac{1}{12}h^2, \quad \beta_1 = -\tfrac{1}{12}h^2.$$

Hence we have the approximate integration formula

$$\int_0^h f(x) \, dx \simeq \tfrac{1}{2}h(f(0) + f(h)) + \tfrac{1}{12}h^2(f'(0) - f'(h)).$$

If $f(x)$ is a polynomial of degree less than or equal to 4, then this formula is exact.

It is not difficult to deduce from this result that

$$\int_{x_0}^{x_n} f(x) \, dx \simeq \tfrac{1}{2}h(f_0 + 2f_1 + 2f_2 + \cdots + 2f_{n-1} + f_n) + \tfrac{1}{12}h^2(f_0' - f_n')$$

in the notation defined earlier. Thus we have obtained the trapezoidal rule with an extra "correction term" $\frac{1}{12}h^2(f_0' - f_n')$.

If the function values have been rounded to k decimal places and the derivative values have been rounded to k_1 decimal places, then the modulus of the round-off error in this method is (compare the trapezoidal rule, page 56) less than or equal to

$$\tfrac{1}{2}(x_n - x_0)10^{-k} + \tfrac{1}{12}h^2 10^{-k_1}.$$

The truncation error can be investigated for twice-differentiable functions using Taylor series.

Example. Use the formula derived above and values of $(1+x)^{1/2}$ correct to two decimal places to obtain an approximation to the integral $\int_2^4 (1+x)^{1/2} dx$.

We have $f(x) = (1+x)^{1/2}$ and so $f'(x) = \dfrac{1}{2(1+x)^{1/2}}$.

Now the modulus of the round-off error in $(1+x)^{1/2}$ is less than or equal to $\frac{1}{2}10^{-2}$ and so the modulus of the error in $\dfrac{1}{2(1+x)^{1/2}}$ is less than or equal to $\dfrac{\frac{1}{2}10^{-2}}{2(1+x)}$ (see §2.6.3, page 16).

Hence the modulus of the error in each of $f'(2)$ and $f'(4)$ is less than or equal to $\frac{1}{6}(\frac{1}{2}10^{-2}) < 0.001$.

Therefore we shall use values of f' to three decimal places. Then with $h = 2$ (which is the largest possible value we can use) the modulus of the round-off error in the formula will be less than or equal to

$$\tfrac{1}{2}(4-2)10^{-2} + \frac{2^2}{12}(0.001) \simeq 0.0103.$$

Note that this error will become even closer to 0.01 as h is decreased. The required values of $(1+x)^{1/2}$ are given in the example on page 61.

x	$(1+x)^{1/2}$	$\dfrac{1}{2(1+x)^{1/2}}$
2	1.73	0.289
4	2.24	0.223

With $h = 2$ we obtain

$$\int_2^4 (1+x)^{1/2}\,dx \simeq \tfrac{1}{2}(2)(1{\cdot}73 + 2{\cdot}24) + \tfrac{1}{12}(2)^2(0{\cdot}289 - 0{\cdot}223)$$

$$= 3{\cdot}992.$$

With $h = 1$ we obtain

$$\int_2^4 (1+x)^{1/2}\,dx \simeq \tfrac{1}{2}(1)(1{\cdot}73 + 2(2{\cdot}00) + 2{\cdot}24) + \tfrac{1}{12}(1)^2(0{\cdot}289 - 0{\cdot}223)$$

$$= 3{\cdot}991$$

rounded to three decimal places.

Hence we now have

$$\int_2^4 (1+x)^{1/2}\,dx = 3{\cdot}99 \pm 0{\cdot}01.$$

EXERCISES

1. Use the method of undetermined coefficients to evaluate (without first changing the origin of x) the coefficients α and β in the integration formula

$$\int_{x_0}^{x_1} f(x)\,dx = \alpha f(x_0) + \beta f(x_1) + E.$$

2. Use the method of undetermined coefficients to evaluate (without first changing the origin of x) the coefficients α, β and γ in the integration formula

$$\int_{x_0}^{x_2} f(x)\,dx = \alpha f(x_0) + \beta f(x_1) + \gamma f(x_2) + E.$$

3. Use the method of undetermined coefficients to evaluate the coefficients a, b, c in the integration formula

$$\int_0^{2h} x^{-1/2} f(x)\,dx = (2h)^{1/2}\{af(0) + bf(h) + cf(2h)\} + E.$$

4. Use the method of undetermined coefficients to evaluate the parameters a and b in the integration formula

$$\int_{-h}^{h} f(x)\,dx = hf(a) + hf(b) + E.$$

Show that the error term E will be zero for polynomials of degree three or less. (This formula is an example of a Gaussian integration formula.)

5. For the integration formula

$$\int_0^h f(x)\,dx \simeq \tfrac{1}{24}h(9f(0) + 19f(h) - 5f(2h) + f(3h))$$

the modulus of the truncation error can be shown to be less than or equal to $\frac{19}{720}h^5 f^{iv}(\xi)$ where $0 \leqslant \xi \leqslant 3h$.

Use this formula to find, correct to three decimal places, $\int_0^{0\cdot2} f(x)dx$ where $f(x)$ is given by the following table and it can be assumed that $|f^{iv}(x)|$ is very much less than 1.

x	0	0·2	0·4	0·6
$f(x)$	1·0000	0·8187	0·6703	0·5488

Iteration

7.1. Introduction

If we require to obtain the roots of the equation $2x^2 - 4x - 3 = 0$, we can make use of a well-known formula and obtain the roots in the form $\frac{1}{4}(4 \pm \sqrt{40})$. However, if we require to obtain the roots of the equation $2x^4 + x^3 - 3x^2 + 4x - 1 = 0$ or of the equation $x = 2 \sin x$, there is no known formula which we can apply to obtain these roots directly. In this chapter we shall consider how to obtain numerical approximations (to any required accuracy) to the roots of such equations. The method employed is called *iteration*.

In an iterative process we start with an approximation x_0 to a root λ and from this obtain another approximation x_1 from which yet another approximation x_2 is obtained and so on. For an effective process (for a particular root) the successive values (or iterates) $x_1, x_2, x_3, \ldots$ should become progressively closer to the root λ. The process is continued until an approximation of the required accuracy is obtained. It is thus clear that for an iterative process we require both

(i) a starting approximation x_0 and

(ii) a method or formula for obtaining the approximation x_{n+1} in terms of the approximation x_n.

7.2. Simple iteration

If the equation whose roots we are attempting to find is $f(x) = 0$, then frequently a starting approximation x_0 can be obtained from a sketch of the graph of $f(x)$.

If the given equation $f(x) = 0$ is rewritten in the form $x = F(x)$, we can obtain the successive iterates $x_1, x_2, x_3, \ldots$ as follows:

$$x_1 = F(x_0),$$
$$x_2 = F(x_1),$$
$$x_3 = F(x_2),$$
$$\ldots \ldots \ldots \ldots$$
$$x_{n+1} = F(x_n),$$
$$\ldots \ldots \ldots \ldots$$

For example, consider the equation

$$2x^3 - 7x + 2 = 0.$$

A rough sketch shows that there is a positive root between 0 and 1 and another between 1 and 2 (Figure 10).

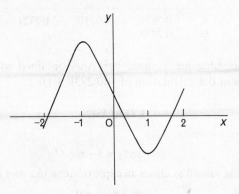

Figure 10

Now the given equation can be rewritten

$$x = \tfrac{2}{7}(x^3 + 1)$$

that is, in the form $x = F(x)$ where $F(x) = \tfrac{2}{7}(x^3 + 1)$. We shall attempt to determine the root between 0 and 1 using the process defined by the starting approximation $x_0 = 1$ and the formula

$$x_{n+1} = \tfrac{2}{7}(x_n^3 + 1).$$

We thus obtain

$$x_1 = \tfrac{2}{7}(1^3 + 1) = 0 \cdot 571 \text{ (rounded to three decimal places)}$$
$$x_2 = \tfrac{2}{7}((0 \cdot 571)^3 + 1) = 0 \cdot 339 \text{ (rounded to three decimal places)}$$

and so on. However the work is better laid out in tabular form as follows:

n	x_n	x_n^3	$\tfrac{2}{7}(x_n^3 + 1)$
0	1	1	0·571
1	0·571	0·186	0·339
2	0·339	0·039	0·297
3	0·297	0·026	0·293
4	0·293	0·025	0·293
5	0·293		

D

The process has converged to 0.293 (rounding to three decimal places). The exact root is in fact $1-\frac{1}{2}\sqrt{2} = 0.292893\ldots$ Greater accuracy in the iterated solution can be obtained by retaining more figures in the calculations. In the following four decimal places have been retained:

n	x_n	x_n^3	$\frac{2}{7}(x_n^3+1)$
5	0.2929	0.0251	0.2929
6	0.2929		

(Note that the value for x_5 used here was obtained by retaining four decimal places in the calculation of $\frac{2}{7}((0.293)^3+1)$.)

EXERCISES

1. Use the iteration formula

$$x_{n+1} = \tfrac{1}{2} - x_n^3$$

and the starting value 0 to obtain an approximation to a root of the equation

$$x^3 + x - \tfrac{1}{2} = 0.$$

Continue the iterations until convergence is obtained using three decimal places.

2. Use the iteration formula

$$x_{n+1} = 1 - \sin x_n$$

and the starting value 1 to obtain an approximation to a root of the equation

$$\sin x + x - 1 = 0.$$

Use three decimal places in the calculations.

As we saw on page 87 the equation $2x^3 - 7x + 2 = 0$ has a second positive root between 1 and 2. We shall now attempt to obtain this root using the same iterative formula as before, namely $x_{n+1} = \frac{2}{7}(x_n^3+1)$ and starting value $x_0 = 2$. We now obtain (rounding the calculations to three decimal places)

n	x_n	x_n^3	$\frac{2}{7}(x_n^3+1)$
0	2	8	2.571
1	2.571	16.994	5.141
2	5.141	135.876	39.107

This process is diverging and will never give the root between 1 and 2 no matter how many iterations are carried out. This illustrates the point that not all iterative processes converge.

The above iteration process is not, however, the only one we can devise for the given equation. The given equation can also be written in the form

$$x = \frac{7x-2}{2x^2} \quad \text{for} \quad x \neq 0,$$

and so we now consider the process defined by the formula

$$x_{n+1} = \frac{7x_n-2}{2x_n^2}$$

and a starting value x_0. With $x_0 = 2$ we obtain (rounding the calculations to three decimal places)

n	x_n	x_n^2	$7x_n-2$	$2x_n^2$	$\dfrac{7x_n-2}{2x_n^2}$
0	2	4	12	8	1·5
1	1·5	2·25	8·5	4·50	1·889
2	1·889	3·568	11·223	7·136	1·573
3	1·573	2·474	9·011	4·948	1·821
......	......	......	......	......	......
22	1·709	2·922	9·963	5·844	1·705
23	1·705	2·908	9·935	5·816	1·708
24	1·708	2·918	9·956	5·836	1·706
25	1·706	2·911	9·942	5·822	1·708
26	1·708				

The process has at last converged although the third decimal place has not been completely determined. It could however be obtained by continuing the iterations retaining more significant figures in the calculations. The exact root is in fact $1 + \frac{1}{2}\sqrt{2} = 1·707106\ldots$.

Before leaving this example it is worth mentioning that the given equation can also be rewritten in the form

$$x = \left(\frac{7x-2}{2x}\right)^{1/2}$$

or in the form

$$x = \left(\frac{7x-2}{2}\right)^{1/3}$$

so that the iterative formulae defined by

$$x_{n+1} = \left(\frac{7x_n-2}{2x_n}\right)^{1/2}$$

and

$$x_{n+1} = \left(\frac{7x_n-2}{2}\right)^{1/3}$$

might also be considered.

A comparison of the use of the different iterative formulae with different starting values is given in the following table:

Iteration formula	Starting value	Result
$x_{n+1} = \frac{2}{7}(x_n^3 + 1)$	1	Converges to the root $0 \cdot 292893 \ldots$
	2	Diverges
$x_{n+1} = \dfrac{7x_n - 2}{2x_n^2}$	1	Converges to the root $1 \cdot 707106 \ldots$
	2	Converges to the root $1 \cdot 707106$
$x_{n+1} = \left(\dfrac{7x_n - 2}{2x_n}\right)^{1/2}$	1	Converges to the root $1 \cdot 707106 \ldots$
	2	Converges to the root $1 \cdot 707106 \ldots$
$x_{n+1} = \left(\dfrac{7x_n - 2}{2}\right)^{1/3}$	1	Converges to the root $1 \cdot 707106 \ldots$
	2	Converges to the root $1 \cdot 707106 \ldots$

7.3. Convergence of iterative processes

As we have seen in the last section, the success of an iterative process for a particular root depends on

(i) the starting value used and
(ii) the iterative formula used.

It would be useful if we could determine, before commencing the iterations, whether or not a particular formula and starting value are likely to yield iterates which converge to the required root. It would be useful also to know which of several converging processes for a given root is likely to converge fastest.

A root λ of the equation $f(x) = 0$ will satisfy $f(\lambda) = 0$ and hence also $\lambda = F(\lambda)$ since this is simply $f(\lambda) = 0$ rewritten in a different form. Thus a root of the given equation is the x-coordinate of the point of intersection of the straight line $y = x$ and the curve $y = F(x)$. Consider Figure 11. These diagrams illustrate different iterative processes geometrically. In each case x_0 is the starting approximation. To obtain the next approximation x_1 first find the intersection of the vertical through the point $(x_0, 0)$ with the curve $y = F(x)$; then find the point of intersection of the horizontal through this point with the straight line $y = x$. The x-coordinate of this latter point of intersection is then $x_1 = F(x_0)$. Similarly x_2 is obtained starting from x_1 and so on. Clearly the processes of Figures a and b are convergent and lead to the root λ. However the processes of Figures c and d are divergent and cannot be used to determine the root λ. What then is the essential difference between the processes of a and b and the processes of c and d? From Figure a we see that the slope of the curve $y = F(x)$ is less than the slope of the line $y = x$. This implies that $F'(x) < 1$

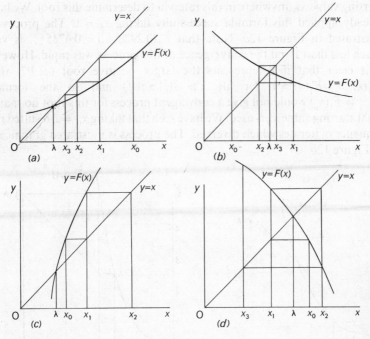

Figure 11

since the slope of the line $y = x$ is equal to 1. From Figure c we see that the slope of the curve $y = F(x)$ is greater than the slope of the line $y = x$. This implies that $F'(x) > 1$. In Figure b, since the slope of the curve $y = F(x)$ is negative, we have that $|F'(x)| < 1$ and in Figure d that $|F'(x)| > 1$. In general it can be shown that a necessary condition for the convergence of the iterative process with formula $x_{n+1} = F(x_n)$ to a root λ is that $|F'(\lambda)| < 1$. Further, it can also be shown that $|F'(x)| < 1$ in an interval I containing a root λ is a sufficient condition for the convergence to λ of the iterative process defined by the formula $x_{n+1} = F(x_n)$ and *any* starting value x_0 in I.

We shall now consider in the light of the above, the formulae devised earlier for use in iterative processes for the roots of the equation $2x^3 - 7x + 2 = 0$.

(i) For the formula $x_{n+1} = \frac{2}{7}(x_n^3 + 1)$ we have that

$$F_1(x) = \tfrac{2}{7}(x^3 + 1) \quad \text{and so} \quad F_1'(x) = \tfrac{6}{7}x^2.$$

Hence for values of x in an interval containing the smaller positive root (e.g. the interval $[0, 1]$) $F_1'(x) < 1$ and so this formula can be used with

starting value x_0 anywhere in this interval to determine this root. We have
already applied this formula successfully taking $x_0 = 1$. The process is
illustrated in Figure 12a. Note that $F_1'(0.2928\ldots) = 0.0735\ldots$ is very
much less than 1 and the convergence of this process was rapid. However
it is clear that if λ represents the larger positive root ($\simeq 1.7$) then
$|F_1'(\lambda)| > 1(F_1'(x) > 1$ for all $x > \sqrt{(\frac{7}{6})} \simeq 1.1)$ and so the formula
$x_{n+1} = F_1(x_n)$ would not give a convergent process for this root no matter
what starting value x_0 is used. We have seen that taking $x_0 = 2$ resulted in a
sequence of iterates which diverged. The process is illustrated graphically
in Figure 12b.

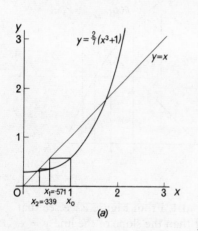

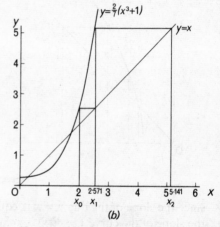

Figure 12

(ii) For the formula $\qquad x_{n+1} = \dfrac{7x_n - 2}{2x_n^2}$

$$F_2(x) = \frac{7x - 2}{2x^2} \quad \text{and} \quad F_2'(x) = \frac{4 - 7x}{2x^3}.$$

It can be deduced that $|F_2'(x)| > 1$ for all x in $[0, \frac{1}{2}]$ and so in particular
for the smaller positive root of the given equation. Hence the formula
$x_{n+1} = F_2(x_n)$ is unsuitable (regardless of starting value x_0) for determining
the smaller root. We have seen that taking $x_0 = 1$ the process did not converge to this root. The process is illustrated in Figure 13a.

It can however be shown that $|F_2'(x)| < 1$ for all x in $[1.5, 2]$ and so for
any starting value x_0 in this interval the formula $x_{n+1} = F_2(x_n)$ will give
a convergent process for the root of the given equation which lies in the

interval $[1·5, 2]$. We have already applied this formula successfully taking $x_0 = 2$. The process is illustrated graphically in Figure 13b.

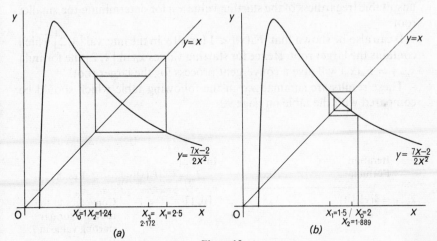

Figure 13

(iii) For the formula $\quad x_{n+1} = \left(\dfrac{7x_n - 2}{2x_n}\right)^{1/2}$

$$F_3(x) = \left(\frac{7x - 2}{2x}\right)^{1/2} \quad \text{and} \quad F_3'(x) = \frac{1}{2x^2}\left(\frac{2x}{7x - 2}\right)^{1/2}.$$

The above formula cannot be applied in the interval $0 < x < \frac{2}{7}$ since it would involve the square root of a negative number. However it can be shown that for all x in the interval $(\frac{2}{7}, \frac{1}{2}]$* which includes the smaller root of the given equation, $|F_3'(x)| > 1$. Thus in particular $|F_3'(x)| > 1$ at the smaller root. Hence the formula $x_{n+1} = F_3(x_n)$ is unsuitable (regardless of the starting value x_0) for determining the smaller root.

It can be shown that $|F_3'(x)| < 1$ for all x in the interval $[\frac{3}{2}, 2]$ which contains the larger root. Hence for starting values in $[\frac{3}{2}, 2]$ the formula $x_{n+1} = F_3(x_n)$ will give a convergent process for the larger root.

(iv) For the formula $\quad x_{n+1} = \left(\dfrac{7x_n - 2}{2}\right)^{1/3}$

$$F_4(x) = \left(\frac{7x - 2}{2}\right)^{1/3} \quad \text{and} \quad F_4'(x) = \frac{7}{6}\left(\frac{2}{7x - 2}\right)^{2/3}.$$

---

* x in the interval $(a, b]$ implies that $a < x \leqslant b$.

It can be shown that $|F_4'(x)| > 1$ for all x in $(\frac{2}{7}, \frac{1}{2}]$ and so in particular is greater than 1 at the smaller root. Hence the formula $x_{n+1} = F_4(x_n)$ is unsuitable (regardless of the starting value x_0) for determining the smaller root.

It can also be shown that $|F_4'(x)| < 1$ for all x in the interval $[\frac{3}{2}, 2]$ which contains the larger root. Hence for starting values x_0 in $[\frac{3}{2}, 2]$ the formula $x_{n+1} = F_4(x_n)$ will give a convergent process for the larger root.

These results are summarized in the following table which should be compared with the table on page 90.

Iteration Formula	$F'(x)$	Interval I	$|F'(x)|$	Results
$x_{n+1} = \frac{2}{7}(x_n^3 + 1)$	$\frac{6}{7}x^2$	$[0, 1]$	< 1	Converges to the root in I for *any* starting value in I.
		$[1 \cdot 1, \infty]$	> 1	Will not converge to the root in I no matter what starting value is used.
$x_{n+1} = \dfrac{7x_n - 2}{2x_n^2}$	$\dfrac{4 - 7x}{2x^3}$	$[0, \frac{1}{2}]$	> 1	Will not converge to the root in I no matter what starting value is used.
		$[\frac{3}{2}, 2]$	< 1	Converges to the root in I for *any* starting value in I.
$x_{n+1} = \left(\dfrac{7x_n - 2}{2x_n}\right)^{1/2}$	$\dfrac{1}{2x^2}\left(\dfrac{2x}{7x - 2}\right)^{1/2}$	$(\frac{2}{7}, \frac{1}{2}]$	> 1	Will not converge to the root in I no matter what starting value is used.
		$[\frac{3}{2}, 2]$	< 1	Converges to the root in I for *any* starting value in I.
$x_{n+1} = \left(\dfrac{7x_n - 2}{2}\right)^{1/3}$	$\dfrac{7}{6}\left(\dfrac{2}{7x - 2}\right)^{2/3}$	$[\frac{2}{7}, \frac{1}{2}]$	> 1	Will not converge to the root in I no matter what starting value is used.
		$[\frac{3}{2}, 2]$	< 1	Converges to the root in I for *any* starting value in I.

Example. Determine an approximation to the only real solution of the equation $f(x) = 0$ in which $f(x) \equiv x^3 - x^2 - x - 3$.

From the table

x	-3	-2	-1	0	1	2	3	4
$f(x)$	-36	-13	-4	-3	-4	-1	12	41

we can obtain a simple sketch of the curve represented by the given equation and from this we see that the required root must lie between $x = 2$ and $x = 3$. This is, of course, clear from the table itself. Since $f(2) = -1$ and $f(3) = 12$ the curve must cross the x-axis somewhere between 2 and 3; that is, there must be a root, λ say, of $f(x) = 0$ in this interval.

Rewriting the equation $f(x) = 0$ in the form

$$x = x^3 - x^2 - 3$$

we consider the formula $x_{n+1} = F_1(x_n)$

where $$F_1(x) = x^3 - x^2 - 3.$$

Then $$F_1'(x) = 3x^2 - 2x.$$

For values of x in the interval $[2, 3]$ the value of $F_1'(x)$ increases steadily from 8 to 21. Hence $|F_1'(x)| > 1$ for all x in $[2, 3]$. Therefore this formula is not suitable for the determination of the root.

Now rewrite the equation $f(x) = 0$ in the form

$$x = 1 + \frac{1}{x} + \frac{3}{x^2}$$

and consider the formula $x_{n+1} = F_2(x_n)$

where $$F_2(x) = 1 + \frac{1}{x} + \frac{3}{x^2}.$$

Then $$F_2'(x) = -\frac{1}{x^2} - \frac{6}{x^3}.$$

For values of x in the interval $[2, 3]$ the value of $F_2'(x)$ increases steadily from -1 to $-\frac{1}{3}$. Hence $|F_2'(x)| < 1$ for all x in the interval $(2, 3]$. Therefore the formula $x_{n+1} = F_2(x_n)$ with starting value x_0 any value in $(2, 3]$ will give a convergent process for the required root λ. However if this root is close to 2, and this could also be deduced from the simple sketch of the curve $f(x) = 0$, then, since $|F_2'(2)| = 1$, we might expect that the convergence of this method would be very slow.

Taking $x_0 = 2.5$ and rounding where necessary to three decimal places we obtain the results in the following table.

n	x_n	$\dfrac{1}{x_n}$	$\dfrac{1}{x_n^2}$	$\left(1+\dfrac{1}{x_n}+\dfrac{3}{x_n^2}\right)$
0	2·5	0·4	0·16	1·88
1	1·88	0·532	0·283	2·381
2	2·381	0·420	0·176	1·948
……	……	……	……	……
23	2·125	0·471	0·221	2·134
24	2·134	0·469	0·220	2·129
25	2·129	0·470	0·221	2·133
26	2·133	0·469	0·220	2·129

(Note that the first and last columns in this table could be omitted.) The process now gets itself into a loop, if the calculations are still rounded to three decimal places, giving alternately the values 2·133 and 2·129. Hence we suspect that the given root lies somewhere between 2·133 and 2·129, that is, that to two decimal places (or three significant figures) the root is 2·13. If more accuracy is required then the process must be continued retaining more figures in the calculations.

Now rewrite the given equation as

$$x = (x^2+x+3)^{1/3}$$

and consider the formula $x_{n+1} = F_3(x_n)$

where $$F_3(x) = (x^2+x+3)^{1/3}.$$

Then $$F_3'(x) = \frac{2x+1}{3(x^2+x+3)^{2/3}}.$$

In the interval from 2 to 3 the numerator of $F_3'(x)$ has maximum value 7 while the denominator has minimum value $3(9)^{2/3}$. Hence the maximum value of

$$F_3'(x) < \frac{7}{3(9)^{2/3}} = \frac{7}{3(81)^{1/3}} < \frac{7}{3(4)} < 1,$$

that is $|F_3'(x)| < 1$ for all x in $[2, 3]$ and so the formula $x_{n+1} = F(x_n)$ with starting value x_0 anywhere in this interval will give a convergent process for the root which lies in this interval.

Also, $$F_3'(2) = \frac{5}{3(9)^{2/3}} \simeq 0.385$$

which is considerably less than $F'_2(2)$ and so, if the required root is close to 2 we might reasonably expect that $|F'_3(\lambda)| < |F'_2(\lambda)|$ so that the process with formula $x_{n+1} = F_3(x_n)$ should converge faster than the process with formula $x_{n+1} = F_2(x_n)$ to the root λ. Taking $x_0 = 2.5$ and rounding where necessary to three decimal places we obtain the results in the following table.

n	x_n	x_n^2	$(x_n^2 + x_n + 3)$
0	2·5	6·25	11·75
1	2·274	5·171	10·445
2	2·186	4·778	9·964
3	2·152	4·631	9·783
4	2·139	4·576	9·715
5	2·134	4·554	9·688
6	2·132	4·546	9·678
7	2·131	4·541	9·672
8	2·131		

Thus, as expected, this method has converged considerably faster than the other method.

A very important point concerning iterative processes must be stated. This is that *small* errors in the calculations are not too important in that they will not *normally* prevent the process from converging to the required root. They may of course result in it being necessary to repeat the iteration process a few more times to obtain convergence. If we are very fortunate, they may even reduce the number of iterations required to obtain convergence! Thus, when carrying out an iterative process, it is not normally necessary to incorporate any checks into the computations. Also, if a result is required to a large number of significant figures, then it is not necessary (and indeed is wasteful of effort if a computer is not being used) to use this number of significant figures in all the calculations right from the beginning. For example, we might start by using only 2 significant figures until the process has converged to this accuracy, and then start using 4 significant figures and later 6 significant figures, and so on.

It is also important to point out that although an iteration process $x_{n+1} = F(x_n)$ with starting value x_0 to determine a value λ may converge, when working to a given number of decimal places, to a value X such that $X = F(X)$ to this number of decimal places, it does *not* follow that X approximates λ to the *same* number of decimal places. Thus, if a root is required to say d decimal places, then the iterative process should be continued until convergence is obtained to at least $(d + 1)$ decimal places.

7.3.1. Flow diagram

A simple flow diagram for carrying out the process of iteration is given below. In this flow diagram t is a measure of the accuracy to which we want our final answer. The value of OLD X which is read in is the starting value x_0 for the iteration. NEW X is computed by evaluating the function F at OLD X.

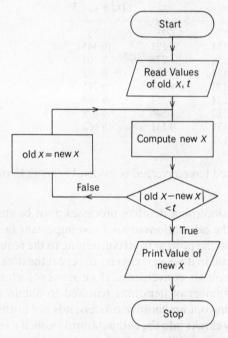

This flow diagram will be quite satisfactory so long as the iterative method being used converges. What will happen if the method does not converge? The reader should modify the flow diagram to take care of the case when the method being used does not converge.

EXERCISES

1. In exercise 1 in the last section we saw that the formula $x_{n+1} = \frac{1}{2} - x_n^3$ with starting value 0 gave a convergent iterative process for a root of the equation $x^3 + x - \frac{1}{2} = 0$. Determine an interval I of the roots such that, using the above formula and any starting value x_0 in I, we can guarantee convergence to the root.

2. Would you expect the formula $x_{n+1} = (\frac{1}{2} - x_n)/x_n^2$ to give a convergent process for the root of the equation in exercise 1 for a sufficiently accurate starting value x_0?

Give reasons for your expectation.

Take $x_0 = 0.4$ and evaluate the next three iterates x_1, x_2, x_3. Work to three decimal places.

(*Hint:* Consider the interval $(0, \frac{1}{2}]$ containing the root).

3. Would you expect the formula $x_{n+1} = (\frac{1}{2} - x_n)^{1/3}$ to give a convergent process for the root of the equation in exercise 1 for a sufficiently accurate starting value x_0?

Give reasons for your expectation.

Take $x_0 = 0.4$ and evaluate the next five iterates x_1, x_2, x_3, x_4, x_5 to three decimal places.

(*Hint:* Consider the interval $[\frac{1}{3}, \frac{1}{2})$ containing the root.)

4. In exercise 2 in the last section we saw that the formula $x_{n+1} = 1 - \sin x_n$ with starting value 1 gave a convergent iterative process for a root of the equation $\sin x + x - 1 = 0$. Determine an interval I of the root such that using the above formula and any starting value x_0 in I we can guarantee convergence to the root.

5. Show that the equation $x^3 - 10x + 4 = 0$ has one root in each of the intervals $[-4, -3]$, $[0, 1]$ and $[2, 3]$.

Show also that the formula $x_{n+1} = \frac{1}{10}(x_n^3 + 4)$ with starting value x_0 chosen in the interval $[0, 1]$ will give a convergent process for the root in this interval.

Is the above formula suitable for determining either of the other two roots?

Obtain an iteration formula which, when applied with any starting value in the interval $[-4, -3]$, will converge to the root in that interval, and which when applied with any starting value in the interval $[2, 3]$ will converge to the root in that interval. Determine each of the three roots correct to 3 significant figures.

6. Show that the equation $\sin x - x + \frac{1}{2} = 0$ has a root in the interval $(0, \frac{1}{2}\pi)$.

Determine the root of the given equation correct to four significant figures.

7.4. Order of iterative processes

We have seen that, if we are given an equation $f(x) = 0$, then it may be possible, by rewriting the equation in the form $x = F(x)$ for different functions F, to obtain several different iterative processes associated with the given equation. We have also seen that some of these methods will converge to a particular root λ of $f(x) = 0$ for sufficiently accurate starting values, while others will not converge to λ, no matter how good a starting value is used. In this section we shall consider in more mathematical detail the convergence of an iterative process, and this will lead us on to consider ways in which the convergence can be speeded up so that it is not necessary to repeat the process so often to obtain a particular root.

If λ is a root of the equation $f(x) = 0$ then $f(\lambda) = 0$. Then, if the iterative formula $x_{n+1} = F(x_n)$ is to be used to determine the root λ, we must have $\lambda = F(\lambda)$.

Let $e_n = x_n - \lambda$, that is $-e_n$ represents the error in the nth iterate x_n.

Then
$$x_n = \lambda + e_n$$

and $$x_{n+1} = \lambda + e_{n+1}.$$

But
$$x_{n+1} = F(x_n) = F(\lambda + e_n)$$
$$= F(\lambda) + e_n F'(\lambda) + \tfrac{1}{2}e_n^2 F''(\lambda) + \cdots,$$

on using Taylor's series expansion.*

Hence
$$\lambda + e_{n+1} = F(\lambda) + e_n F'(\lambda) + \tfrac{1}{2}e_n^2 F''(\lambda) + \cdots$$
$$= \lambda + e_n F'(\lambda) + \tfrac{1}{2}e_n^2 F''(\lambda) + \cdots.$$

Therefore
$$e_{n+1} = e_n F'(\lambda) + \tfrac{1}{2}e_n^2 F''(\lambda) + \cdots.$$

Now, if the process is convergent, there must be a stage in the iteration beyond which each successive error e_{n+1} must be numerically less than the modulus of the previous error e_n; that is, $|e_{n+1}| < |e_n|$ and the successive iterates are getting closer and closer to the value of the root λ.

Then, when n is sufficiently large to make e_n "small", we have

$$e_{n+1} \simeq e_n F'(\lambda) \quad \text{if} \quad F'(\lambda) \neq 0.$$

In this case the process is said to be *first order*. Since for such a process, once n is large enough, an approximation to the error at any stage is obtained by multiplying the error at the previous stage by the factor $F'(\lambda)$, it follows that a necessary condition for convergence is that $|F'(\lambda)| < 1$. Also, the rate of convergence will increase as $|F'(\lambda)|$ decreases. Hence if $|F'(\lambda)|$ is just less than 1, the convergence will be very slow and it will be necessary to repeat the process many times to obtain an accurate value of λ. If $|F'(\lambda)|$ is very much less than one, the convergence will be faster, and it should not be necessary to repeat the process so often in order to obtain the same final accuracy.

If $F'(\lambda) = 0$ then we have

$$e_{n+1} \simeq \tfrac{1}{2}e_n^2 F''(\lambda) \quad \text{if} \quad F''(\lambda) \neq 0.$$

In this case the process is said to be *second order*. For such a process an approximation to the error at any stage is obtained by *squaring the error* at the previous stage and then multiplying by the factor $\tfrac{1}{2}F''(\lambda)$. Thus, once we have got close enough to the desired root, so that the error in our approximation is "small" (that is, less than one unit on some scale of measurement) the errors in the subsequent iterates will decrease very rapidly.

In a similar way, third and higher-order processes can be defined.

---

* See Appendix 1.

Now first-order processes have the advantage that they are generally associated with very simple formulae. Their disadvantage, however, is that their convergence will be relatively slow (compared with higher-order processes) so that, although the work involved in carrying out one step of the iteration may be quite small, it will generally be necessary to repeat this many times. Second-order processes on the other hand will generally involve more complicated formulae, so that the work involved for each step of the iteration is increased. Their faster convergence rates, however, mean that this will not have to be repeated so often. The formulae for third and higher-order methods are generally so complicated that the extra gain due to their even-faster convergence rates is more than offset by the added labour of computation at each step. Thus, in general, only first and second-order methods are widely used in practice.

Example. Consider the methods which have already been found to converge (for suitable starting values) to the positive roots of the equation $2x^3 - 7x + 2 = 0$.

Let λ_1 be the root between $x = 0$ and $x = 1$ and let λ_2 be the root between $x = 1$ and $x = 2$.

(i) The method $\qquad x_{n+1} = \tfrac{2}{7}(x_n^3 + 1) = F_1(x_n)$

converges, for suitable starting values, to λ_1.

$$F_1'(\lambda_1) = \tfrac{6}{7}\lambda_1^2 \neq 0 \quad \text{since} \quad \lambda_1 \neq 0.$$

Hence the method is first order for the root λ_1.

(ii) The method $\qquad x_{n+1} = \dfrac{7x_n - 2}{2x_n^2} = F_2(x_n)$

converges, for suitable starting values, to λ_2.

$$F_2'(\lambda_2) = \frac{4 - 7\lambda_2}{2\lambda_2^3} \neq 0 \quad \text{if} \quad \lambda_2 \neq \tfrac{4}{7}.$$

Hence the method is first order for the root $\lambda_2(\neq \tfrac{4}{7})$.

(iii) The method $\qquad x_{n+1} = \left(\dfrac{7x_n - 2}{2x_n}\right)^{1/2} = F_3(x_n)$

converges, for suitable starting values, to λ_2.

$$F_3'(\lambda_2) = \frac{1}{2\lambda_2^2}\left(\frac{2\lambda_2}{7\lambda_2 - 2}\right)^{1/2} = \frac{1}{(2\lambda_2^3(7\lambda_2 - 2))^{1/2}} \neq 0.$$

Hence the method is first order for the root λ_2.

(iv) The method $\quad x_{n+1} = \left(\dfrac{7x_n - 2}{2}\right)^{1/3} = F_4(x_n)$

converges, for suitable starting values, to λ_2.

$$F_4'(\lambda_2) = \frac{7}{6}\left(\frac{2}{7\lambda_2 - 2}\right)^{2/3} \neq 0.$$

Hence the method is first order for the root λ_2.

Example. Assuming that the iterative process defined by the formula $x_{n+1} = \frac{1}{2}(x_n + a/x_n)$ in which a is a positive constant, and a given starting value x_0 converges, what will be the limiting value of x_n as n becomes very large?

Show that the method defined above is second order.

If the process converges to the value λ, then we must have

$$\lambda = \tfrac{1}{2}\left(\lambda + \frac{a}{\lambda}\right)$$

(that is $\lambda = F(\lambda)$ in the notation of this chapter).

Hence $\qquad\qquad\qquad 2\lambda = \lambda + \dfrac{a}{\lambda}$

and so $\qquad\qquad\qquad \lambda^2 = a.$

Thus the process will converge to a value λ which is the square root of a.

Now $\qquad\qquad F(x) = \tfrac{1}{2}\left(x + \frac{a}{x}\right).$

Therefore $\qquad\qquad F'(x) = \tfrac{1}{2}\left(1 - \frac{a}{x^2}\right)$

and so $\quad F'(\lambda) = \tfrac{1}{2}\left(1 - \frac{a}{\lambda^2}\right) = 0 \quad$ since $\quad \lambda^2 = a.$

Hence the given method is at least second order.

$$F''(x) = \frac{a}{x^3}$$

and so $\qquad F''(\lambda) = \dfrac{a}{\lambda^3} \neq 0 \quad$ since a is positive.

Hence the given method is second order.

EXERCISES

1. Show that the following convergent iterative processes are both first order:
  (i) $x_{n+1} = \frac{1}{2} - x_n^3$ with $x_0 \in [0, \frac{1}{2}]$ for a root of $x^3 + x - \frac{1}{2} = 0$,
  (ii) $x_{n+1} = 1 - \sin x_n$ with $x_0 \in (0, \pi)$ for a root of $\sin x + x - 1 = 0$.

2. Show that the following convergent iterative processes for roots of the equation $x^3 - 10x + 4 = 0$ are first order
  (i) $x_{n+1} = \frac{1}{10}(x_n^3 + 3)$ with $x_0 \in [0, 1]$,
  (ii) $x_{n+1} = (10x_n - 4)^{1/3}$ with $x_0 \in [2, 3]$.

3. The iterative formula $x_{n+1} = \frac{1}{2}x_n(3 - ax_n^2)$ can be used to determine $1/\sqrt{a}$. Show that the process is second order and apply it to obtain $1/\sqrt{7}$ correct to six decimal places. (Check your result against that obtained from tables.)

7.5. The rule of false position

Once again consider the determination of a root of the equation

$$f(x) = 0.$$

The rule of false position is best explained geometrically. In Figure 14 is sketched a part of the graph of the curve represented by the equation $y = f(x)$. The root of $f(x) = 0$ which we are seeking is represented by the x-coordinate of the point P, the intersection of the curve represented by $y = f(x)$ with the x-axis. Using the rule of false position we start with two points $A(\alpha, f(\alpha))$ and $B_0(x_0, f(x_0))$ which are preferably on opposite sides of the x-axis and determine the x-coordinate, x_1, of the point P_1 which is the point of intersection of the x-axis and the straight line AB_0.

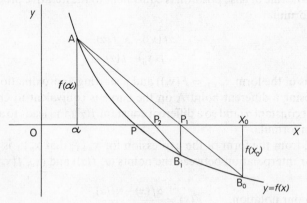

Figure 14

Now, starting from points A and $B_1(x_1, f(x_1))$ (on the curve $y = f(x)$), we can determine x_2, the x-coordinate of the point P_2 the point of intersection of the x-axis and the straight line AB_1. Thus we can generate a sequence of values $x_0, x_1, x_2, \ldots$ which we hope will converge to the x-coordinate of the point P, that is, to the required root of the given equation.

The method is, of course, not applied graphically but analytically. An equation for the line AB_0 is

$$y - f(\alpha) = \frac{f(x_0) - f(\alpha)}{x_0 - \alpha}(x - \alpha).$$

Now this passes through P_1 the point $(x_1, 0)$

and so

$$-f(\alpha) = \frac{f(x_0) - f(\alpha)}{x_0 - \alpha}(x_1 - \alpha).$$

Hence

$$x_1 = \frac{\alpha f(x_0) - x_0 f(\alpha)}{f(x_0) - f(\alpha)}.$$

Similarly

$$x_2 = \frac{\alpha f(x_1) - x_0 f(\alpha)}{f(x_1) - f(\alpha)}$$

and in general, for $n \geqslant 0$, we have

$$x_{n+1} = \frac{\alpha f(x_n) - x_n f(\alpha)}{f(x_n) - f(\alpha)}.$$

Now regarding the point A as fixed, and so α and $f(\alpha)$ as constants, we see that the rule of false position is equivalent to the iterative process given by the formula

$$x_{n+1} = \frac{\alpha f(x_n) - x_n f(\alpha)}{f(x_n) - f(\alpha)}$$

(which is of the form $x_{n+1} = F(x_n)$) and a starting approximation x_0.

Choosing a different point A on the curve is equivalent to choosing a different constant α (and so a different constant $f(\alpha)$) and leads to a different iteration formula.

Note, from the form of the expression for x_{n+1}, that x_{n+1} is obtained by linear interpolation between the points $(\alpha, f(\alpha))$ and $(x_n, f(x_n))$.

Now, in our notation, $F(x) = \dfrac{\alpha f(x) - x f(\alpha)}{f(x) - f(\alpha)}$

and so $F'(x) = \dfrac{(f(x) - f(\alpha))(\alpha f'(x) - f(\alpha)) - (\alpha f(x) - x f(\alpha))f'(x)}{(f(x) - f(\alpha))^2}.$

Hence if λ is a root of $f(x) = 0$ (and so $f(\lambda) = 0$),

$$F'(\lambda) = \frac{f(\alpha)(f(\alpha) - \alpha f'(\lambda) + \lambda f'(\lambda))}{(f(\alpha))^2}$$

$$= \frac{f(\alpha) - \alpha f'(\lambda) + \lambda f'(\lambda)}{f(\alpha)}$$

if $f(\alpha) \neq 0$.

Therefore, in general, $F'(\lambda) \neq 0$ and so the rule of false position is equivalent to a first-order iterative method.

Example. Use the rule of false position to determine approximations to the positive roots of the equation

$$2x^3 - 7x + 2 = 0.$$

(Use three decimal places in the calculations.)

Let $f(x) \equiv 2x^3 - 7x + 2$.

We have seen already that there is a root of $f(x) = 0$ between $x = 0$ and $x = 1$. Thus we shall take $\alpha = 0$ and $x_0 = 1$. Then, since $f(0) = 2$, the formula for the iterative method corresponding to the rule of false position (with this choice of α) is

$$x_{n+1} = \frac{-2x_n}{f(x_n) - 2} = \frac{-2x_n}{2x_n^3 - 7x_n}$$

$$= \frac{-2}{2x_n^2 - 7}$$

if $x_n \neq 0$.

Hence in our notation $F(x) = \dfrac{-2}{2x^2 - 7}$

and so $F'(x) = \dfrac{8x}{(2x^2 - 7)^2}.$

Now $8x$ increases steadily from 0 to 8 and $(2x^2 - 7)^2$ decreases steadily from 49 to 25 as x increases from 0 to 1. Thus the maximum value of $F'(x)$ for x in $[0, 1]$ occurs at $x = 1$ and is $\frac{8}{25}$. Therefore $|F'(x)| < 1$ for x in $[0, 1]$ and so the iterative process is convergent for any starting value x_0 in $[0, 1]$ and in particular for $x_0 = 1$.

With $x_0 = 1$ the calculation proceeds as follows:

n	x_n	$2x_n^2$	$2x_n^2 - 7$
0	1·000	2·000	$-5\cdot000$
1	0·400	0·320	$-6\cdot680$
2	0·299	0·179	$-6\cdot821$
3	0·293	0·172	$-6\cdot828$
4	0·293		

Hence the process has converged (rounding to three decimal places). For a better approximation to the root (whose exact value is $0\cdot292893\ldots$) the process must be continued retaining more significant figures in the calculations.

Since the second positive root is known to lie between 1 and 2, let us consider the process obtained by taking $\alpha = 2$ and $x_0 = 1$. Since $f(2) = 4$ the corresponding iteration formula is

$$x_{n+1} = \frac{2f(x_n) - 4x_n}{f(x_n) - 4}$$

$$= \frac{4x_n^3 - 18x_n + 4}{2x_n^3 - 7x_n - 2}$$

$$= \frac{(x_n - 2)(4x_n^2 + 8x_n - 2)}{(x_n - 2)(2x_n^2 + 4x_n + 1)}$$

$$= \frac{2\{2x_n(x_n + 2) - 1\}}{2x_n(x_n + 2) + 1}$$

if $x_n \neq 2$.

Now
$$F(x) = \frac{2\{2x(x + 2) - 1\}}{2x(x + 2) + 1}$$

and so, after differentiating and simplifying we obtain

$$F'(x) = \frac{16(x + 1)}{(2x^2 + 4x + 1)^2}.$$

Now for x in $[1, 2]$ the maximum value of the numerator occurs at $x = 2$ and is 48 and the minimum value of the denominator occurs at $x = 1$ and is 49. Hence the maximum value of $F'(x)$ for x in $[1, 2]$ is certainly less than $\frac{48}{49}$. Thus $|F'(x)| < 1$ for x in $[1, 2]$ and so the process is convergent for any starting value x_0 in this interval and in particular for $x_0 = 1$.

With $x_0 = 1$ the calculation proceeds as follows:

n	x_n	$2x_n(x_n+2)$
0	1·000	6·000
1	1·429	9·800
2	1·630	11·834
3	1·688	12·451
4	1·703	12·612
5	1·706	12·645
6	1·707	12·656
7	1·707	

Hence the process has converged (rounding to three decimal places). For a better approximation to the root the process can be continued retaining more significant figures in the calculations.

EXERCISES

1. Use the rule of false position to determine the positive roots of the equation $x^3 - 10x + 4 = 0$ correct to three significant figures. (Note, we have already seen in exercise 5, page 99, that the positive roots lie in the intervals $[0, 1]$, $[2, 3]$.)

2. Use the rule of false position to determine the root of the equation $\sin x + x - 1 = 0$ correct to two significant figures. (Note it is easy to see that the root lies in $[0, \pi]$.)

7.6. Aitken's δ^2-process

We have seen that using simple (or first-order) iteration convergence can be very slow indeed and that using second- or higher-order (and generally more complicated) methods should result in faster convergence. Before considering second-order processes, however, we will look at a method by which the convergence of a first-order method can be speeded up. The method, known as Aitken's δ^2-process, is of wide applicability and can be used to speed the convergence of any (convergent) sequence of numbers and not simply those obtained from iterative processes.

For a convergent first-order process $x_{n+1} = F(x_n)$ for the root λ we have as in section 7.4,

$$e_{n+1} = F'(\lambda)e_n + \tfrac{1}{2}F''(\lambda)e_n^2 + \dots$$
$$= e_n(A + E_n)$$

where $A = F'(\lambda)$ and $E_n = \tfrac{1}{2}F''(\lambda)e_n + \dots$ and E_n can be made as small as we please by making n large enough. We cannot calculate the constant A, as λ is as yet unknown.

Then, in the same notation, $e_{n+2} = e_{n+1}(A + E_{n+1})$.
Hence, eliminating A, we obtain

$$\frac{e_{n+1}}{e_n} - \frac{e_{n+2}}{e_{n+1}} = E_n - E_{n+1} \simeq 0$$

for suitably large values of n.

Therefore $\qquad\qquad e_{n+1}^2 - e_n e_{n+2} \simeq 0$

and so $\qquad (\lambda - x_{n+1})^2 - (\lambda - x_n)(\lambda - x_{n+2}) \simeq 0.$

Therefore $\qquad \lambda(x_{n+2} - 2x_{n+1} + x_n) \simeq x_n x_{n+2} - x_{n+1}^2$

and so $\qquad\qquad \lambda \simeq \dfrac{x_n x_{n+2} - x_{n+1}^2}{x_{n+2} - 2x_{n+1} + x_n}.$

This estimate of λ will be better than any of x_n, x_{n+1}, x_{n+2}. Now it is easily verified that this estimate can also be written

$$x_n - \frac{(x_{n+1} - x_n)^2}{x_{n+2} - 2x_{n+1} + x_n}$$

The numbers $(x_{n+2} - x_{n+1})$ and $(x_{n+1} - x_n)$ are the first central differences $\delta x_{n+3/2}$ and $\delta x_{n+1/2}$ respectively of x. Then the number

$$\{(x_{n+2} - x_{n+1}) - (x_{n+1} - x_n)\} = \{\delta x_{n+3/2} - \delta x_{n+1/2}\}$$

is the second central difference $\delta^2 x_{n+1}$ of x. Thus the above result can now be written

$$\lambda \simeq x_n - \frac{(\delta x_{n+1/2})^2}{(\delta^2 x_{n+1})}.$$

Hence the name "δ^2-process". The method of application is as follows:

From a starting value x_0 compute the first iterate $x_1 = F(x_0)$ and thence the second iterate $x_2 = F(x_1)$. Now calculate $x_0^{(1)}$ where

$$x_0^{(1)} = \frac{x_2 x_0 - x_1^2}{x_2 + x_0 - 2x_1} = x_0 - \frac{(\delta x_{1/2})^2}{\delta^2 x_1}.$$

(The best way of calculating this is by first calculating the differences $\delta x_{3/2}$ and $\delta x_{1/2}$ and then the difference $\delta^2 x_1$.)

Now, using $x_0^{(1)}$ as a starting value, calculate the next two iterates $x_1^{(1)} = F(x_0^{(1)})$ and $x_2^{(1)} = F(x_1^{(1)})$. Then calculate

$$x_0^{(2)} = x_0^{(1)} - \frac{(\delta x_{1/2}^{(1)})^2}{\delta^2 x_1^{(1)}}$$

where $\delta x_{1/2}^{(1)} = x_1^{(1)} - x_0^{(1)}$ and $\delta^2 x_1^{(1)} = (x_2^{(1)} - x_1^{(1)}) - (x_1^{(1)} - x_0^{(1)})$.

In this way generate the sequence $\{x_0^{(n)}\}$. This sequence will then tend to the root λ faster than the (first-order) sequence $\{x_n\}$. It can, in fact, be shown that using the δ^2-process along with a first-order method is equivalent to using a second-order method. The calculations are conveniently recorded in tabular form as illustrated in the following example.

Example. Consider again the determination of an approximation to the real solution of the equation $x^3 - x^2 - x - 3 = 0$ using the method

$$x_{n+1} = 1 + \frac{1}{x_n} + \frac{3}{x_n^2}$$

and Aitken's δ^2-process (see page 95).

As before we shall start with $x_0 = 2.5$ and round-off to three decimal places. We then obtain:

x	δx	$\delta^2 x$	$\dfrac{(\delta x)^2}{\delta^2 x}$	$x_0^{(i)}$
2·5			0·343	2·157
1·88	−0·62	1·121		
2·381	0·501			
2·157			0·026	2·131
2·108	−0·049	0·091		
2·150	0·042			
2·131				
2·130				
2·131				

Hence the result 2·13 to two decimal places has been obtained *very* much faster than when using the same method without the δ^2-process.

Using the δ^2-process along with the method $x_{n+1} = (x_n^2 + x_n + 3)^{1/3}$ and again taking $x_0 = 2.5$ we obtain:

x	δx	$\delta^2 x$	$\dfrac{(\delta x)^2}{\delta^2 x}$	$x_0^{(i)}$
2·5			0·370	2·130
2·274	−0·226	0·138		
2·186	−0·088			
2·130				
2·130				

Once again the result 2·13 to two decimal places has been obtained faster than when using the same method without the δ^2-process.

Example. Determine an approximation to the positive root of the equation $f(x) = 0$ where $f(x) = x - 2 \sin x$, that is, of the equation $x - 2 \sin x = 0$ (in which x is measured in radians).

If rough graphs of the curves $y = x$ and $y = 2 \sin x$ are drawn (Figure 15) we readily observe that they intersect, and so $x = 2 \sin x$ (or $x - 2 \sin x = 0$) for a value of x greater than $\frac{1}{2}\pi$. But since the given equation implies that $\sin x = \frac{1}{2}x$ and $\sin x$ is less than or equal to 1, we see that the root must be less than or equal to 2. Hence, if λ is the required root, then $\frac{1}{2}\pi < \lambda \leqslant 2$.

Now the given equation can be rewritten in the form $x = 2 \sin x$ and so we shall consider the iterative method given by $x_{n+1} = F(x_n)$ where $F(x) = 2 \sin x$. Then $F'(x) = 2 \cos x$ and so $F'(\lambda) \neq 0$ since $\frac{1}{2}\pi < \lambda \leqslant 2$. Therefore the above iterative method is first order. Also $|F'(x)| < 1$ for $\frac{1}{2}\pi < x \leqslant 2$ and so we have a neighbourhood of the root for which $|F'(x)| < 1$. Hence the above iterative process will converge to the root for sufficiently accurate starting values. Taking $x_0 = 2$ and using Aitken's δ^2-process along with the above iterative method we obtain the following results rounded to three decimal places:

x	δx	$\delta^2 x$	$\dfrac{(\delta x)^2}{\delta^2 x}$	$x_0^{(i)}$
2			0·109	1·891
$2 \sin 2 = 1\cdot819$	$-0\cdot181$			
	$0\cdot120$	$0\cdot301$		
$2 \sin 1\cdot819 = 1\cdot939$				
1·891			$-0\cdot004$	1·895
$2 \sin 1\cdot891 = 1\cdot898$	$0\cdot007$			
	$-0\cdot004$	$-0\cdot011$		
$2 \sin 1\cdot898 = 1\cdot894$				
1·895				
$2 \sin 1\cdot895 = 1\cdot896$				
$2 \sin 1\cdot896 = 1\cdot895$				

Hence we suspect that the required root lies somewhere between 1·895 and 1·896. To three-figure accuracy then we have $\lambda \simeq 1\cdot90$. Greater accuracy can, of course, be obtained by continuing the process and retaining more significant figures in the calculations. (This would require the use of better tables.)

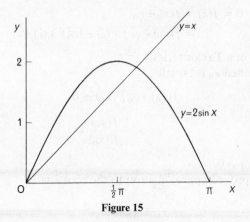

Figure 15

EXERCISES

1. Use the iterative formula $x_{n+1} = \frac{1}{2} - x_n^3$ with starting value 0 and Aitken's δ^2-process to obtain the real root of the equation $x^3 + x - \frac{1}{2} = 0$ correct to three decimal places.

2. Use the iteration formulae $x_{n+1} = 1 - \sin x_n$ with starting value 1 and Aitken's δ^2-process to obtain the real root of the equation $\sin x + x - 1 = 0$ correct to two decimal places.

3. Use the iteration formula $x_{n+1} = (10x_n - 4)^{1/3}$ with starting value 2 and Aitken's δ^2-process to obtain the root of the equation $x^3 - 10x + 4 = 0$ which lies in the interval $[2, 3]$ correct to two decimal places.

4. Use the iteration formula $x_{n+1} = \frac{1}{10}(x_n^3 + 4)$ with starting value 0 and Aitken's δ^2-process to obtain the root of the equation $x^3 - 10x + 4 = 0$ which lies in the interval $[0, 1]$ correct to three decimal places.

5. In the notation used in the text, show that

$$x_0^{(1)} = x_2 - \frac{(x_{n+2} - x_{n+1})^2}{x_{n+2} - 2x_{n+1} + x_n}$$

$$= x_2 - \frac{(\delta x_{n+3/2})^2}{\delta^2 x_{n+1}}.$$

6. Construct a flow diagram for Aitken's δ^2-process.

7.7. The Newton-Raphson Method

In this section we shall consider a very important iterative method which is, in general, second order.

Let x_0 be an approximation to the root λ of the equation $f(x) = 0$ and let e_0 be the error in this approximation so that

$$\lambda \doteq x_0 + e_0.$$

Then $\qquad 0 = f(\lambda) = f(x_0 + e_0)$
$$= f(x_0) + e_0 f'(x_0) + \tfrac{1}{2} e_0^2 f''(x_0) + \ldots$$

on expanding in a Taylor series.*

Hence, provided e_0 is "small",

$$f(x_0) + e_0 f'(x_0) \simeq 0$$

and so $\qquad\qquad\qquad\qquad e_0 \simeq -\dfrac{f(x_0)}{f'(x_0)}$

provided $f'(x_0) \neq 0$.

Thus an improved approximation to the root λ is

$$x_1 = x_0 - \frac{f(x_0)}{f'(x_0)}$$

and then a further improved approximation is

$$x_2 = x_1 - \frac{f(x_1)}{f'(x_1)}$$

provided $e_1 \simeq -\dfrac{f(x_1)}{f'(x_1)}$ is "small".

Hence we have the iterative formula given by

$$x_{n+1} = x_n - \frac{f(x_n)}{f'(x_n)}$$

or $\qquad\qquad\qquad\qquad x_{n+1} = F(x_n)$

where $\qquad\qquad\qquad\qquad F(x) = x - \dfrac{f(x)}{f'(x)},$

7.7.1. Geometrical interpretation

In Figure 16 $OA_0 = x_0$ and $B_0 \equiv (x_0, f(x_0))$ so that $A_0 B_0 = f(x_0)$. $A_1 B_0$ is the tangent to the curve $y = f(x)$ at the point B_0 on it. Therefore the gradient of the line $A_1 B_0$ is $f'(x_0)$.

Hence $\qquad\qquad\qquad f'(x_0) = \dfrac{A_0 B_0}{A_1 A_0} = \dfrac{f(x_0)}{A_1 A_0}$

and so $\qquad\qquad\qquad A_1 A_0 = \dfrac{f(x_0)}{f'(x_0)}.$

---

* See Appendix 1.

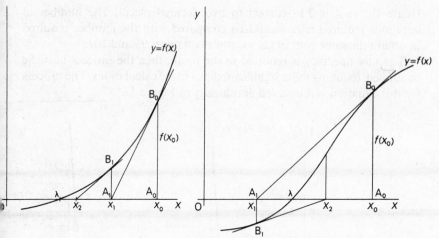

Figure 16

Therefore $\quad x_1 = \overset{\bullet}{x}_0 - \dfrac{f(x_0)}{f'(x_0)} = OA_0 - A_1 A_0 = OA_1$

that is, x_1 is obtained by determining the intersection with the x-axis of the tangent to the curve $y = f(x)$ at the point $(x_0, f(x_0))$ on it. x_2 is obtained in a similar way, starting from the point $(x_1, f(x_1))$. In general, x_{n+1} is obtained by determining the intersection with the x-axis of the tangent to the curve $y = f(x)$ at the point $(x_n, f(x_n))$ on it.

Example. Use the Newton-Raphson method to determine an approximation to the real root of the equation $x^3 - x^2 - x - 3 = 0$.

Using the above notation we have $f(x) \equiv x^3 - x^2 - x - 3$ and so the Newton-Raphson method is given by

$$x_{n+1} = x_n - \frac{f(x_n)}{f'(x_n)} = x_n - \frac{x_n^3 - x_n^2 - x_n - 3}{3x_n^2 - 2x_n - 1}.$$

Starting with $x_0 = 2{\cdot}5$ (as in the example on pages 95 and 109) and rounding to three decimal places throughout we obtain the following results:

x	$f(x)$	$f'(x)$	$\dfrac{f(x)}{f'(x)}$
2·5	3·875	12·75	0·304
2·196	0·571	9·075	0·063
2·133	0·022	8·383	0·003
2·130	−0·003	8·351	0·000

Hence the root is 2·13 correct to two decimal places. The number of iterations required here should be compared with the number required to obtain the same root in the examples on pages 95 and 109.

If greater accuracy is required in the result, then the process must be continued, retaining more significant digits in the calculations. The process for this equation is illustrated graphically in Figure 17.

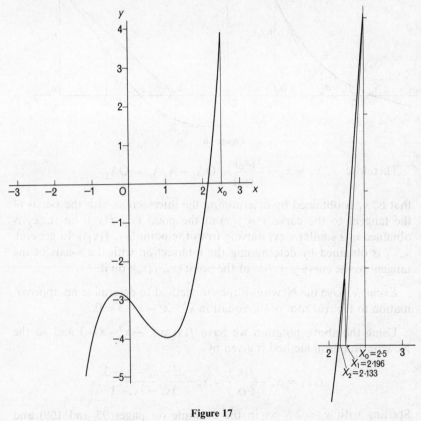

Figure 17

In the above example all the calculations were carried out to three decimal places. Clearly this was a little wasteful of effort since, for example at the last step, there was no need to calculate $f'(x)$ to four significant figures (that is, in this case three decimal places) when $f(x)$ was only known correct to one significant digit. One significant figure would have been sufficient for $f'(x)$. Also, at the previous step, it would have been sufficient to have evaluated $f'(x)$ correct to only one decimal place, or two significant figures since $f(x)$ has only been evaluated to two significant figures.

Example. Use the Newton-Raphson method to determine an approximation to the positive root of the equation $x - 2 \sin x = 0$.

Here the method is given by

$$x_{n+1} = x_n - \frac{x_n - 2 \sin x_n}{1 - 2 \cos x_n} = x_n - \frac{f(x_n)}{f'(x_n)}.$$

Starting with $x_0 = 2$, we obtain the following results:

x	$2 \sin x$	$2 \cos x$	$f(x)$	$f'(x)$	$\dfrac{f(x)}{f'(x)}$
2	1·819	−0·832	0·181	1·832	0·099
1·901	1·892	−0·648	0·009	2	0·004
1·897	1·894	−0·6	0·003	2	0·002
1·895	1·896	−0·6	−0·001	2	

Hence (as before) we obtain the value 1·90 correct to two decimal places for the required root. If greater accuracy is required, the process must be continued, retaining more figures in the calculations.

7.7.2. *Convergence of the process*

Like other iterative processes the Newton-Raphson process has the form $x_{n+1} = F(x_n)$ and will not always converge. For this method

$$F(x) = x - \frac{f(x)}{f'(x)}$$

and so

$$F'(x) = 1 - \frac{f'(x)f'(x) - f(x)f''(x)}{[f'(x)]^2} = \frac{f(x)f''(x)}{[f'(x)]^2}.$$

Hence for convergence we require that

$$\left| \frac{f(x)f''(x)}{[f'(x)]^2} \right| < 1$$

for x in an interval containing the root λ. In practice, however, it is often difficult to determine an interval satisfying this condition, and the following set of conditions can be shown to be sufficient to ensure convergence for any starting value x_0 in an interval $a \leqslant x \leqslant b$.

(i) $f(a)f(b) < 0$.

This implies that $f(a)$ and $f(b)$ have different signs so that the curve represented by the equation $y = f(x)$ crosses the x-axis at least once in the given interval, and so the given equation has at least one root in this interval.

(ii) $f'(x) \neq 0$ for $a \leqslant x \leqslant b$.

This implies that there are no stationary points of the curve $y = f(x)$ in the given interval and so there is only one root of the equation $f(x) = 0$ in this interval.

(iii) $f''(x)$ is either $\geqslant 0$ or $\leqslant 0$ for all x such that $a \leqslant x \leqslant b$.

Thus $f''(x)$ does not change sign in the given interval and so the curve represented by the equation $y = f(x)$ is either concave up over the whole interval or concave down over the whole interval.

(iv) $\left|\dfrac{f(c)}{f'(c)}\right| \leqslant b - a$ where c denotes a if $|f'(a)| < |f'(b)|$ and c denotes b if

$|f'(b)| < |f'(a)|$.

This implies that the tangent to the curve represented by the equation $y = f(x)$ at the end of the interval $a \leqslant x \leqslant b$ at which $|f'(x)|$ is smallest intersects the x-axis within the above interval, so that successive iterates always lie within this interval.

Example. Consider the application of the Newton-Raphson method to determine (a) the root of $x^3 - x^2 - x - 3 = 0$ which lies between 2 and 3 (see the example on page 113) and (b) the root of $x - 2 \sin x = 0$ which lies between $\frac{1}{2}\pi$ and 2 (see the example on page 115).

(a) We have $f(x) = x^3 - x^2 - x - 3$.
   (i) $f(2) = -1$, $f(3) = 12$ and so $f(2)f(3) < 0$.
   (ii) $f'(x) = 3x^2 - 2x - 1 = (3x + 1)(x - 1)$ and so $f'(x) \neq 0$ for $2 \leqslant x \leqslant 3$.
   (iii) $f''(x) = 6x - 2 > 0$ for all x such that $2 \leqslant x \leqslant 3$.
   (iv) $|f'(2)| = 7$ and $|f'(3)| = 20$.

Therefore $\qquad\qquad\qquad |f'(2)| < |f'(3)|$.

Then $\qquad\qquad\qquad \left|\dfrac{f(2)}{f'(2)}\right| = \dfrac{1}{7} < 3 - 2$.

Hence convergence of the Newton-Raphson process is guaranteed for any starting value x_0 such that $2 \leqslant x_0 \leqslant 3$. The results obtained with $x_0 = 2 \cdot 5$ are shown on page 113.

(b) We have $f(x) = x - 2 \sin x$.

   (i) $f(\frac{1}{2}\pi) = \frac{1}{2}\pi - 2 \sin \frac{1}{2}\pi = \frac{1}{2}\pi - 2 < 0$.
      $f(2) = 2 - 2 \sin 2 > 0$ since $\sin 2 < 1$.

      Therefore $\qquad\qquad\qquad\qquad f(\frac{1}{2}\pi) f(2) < 0$.

  (ii) $f'(x) = 1 - 2 \cos x \neq 0$ for $\frac{1}{2}\pi \leqslant x \leqslant 2$ since $\cos x \leqslant 0$ for x in this interval.

 (iii) $f''(x) = 2 \sin x > 0$ for all x such that $\frac{1}{2}\pi \leqslant x \leqslant 2$.

 (iv) $f'(\frac{1}{2}\pi) = 1 - 2 \cos \frac{1}{2}\pi = 1$ and $f'(2) = 1 - 2 \cos 2 > 1$ since $\cos 2 < 0$.

      Then $\qquad\qquad\qquad \left| \dfrac{f(\frac{1}{2}\pi)}{f'(\frac{1}{2}\pi)} \right| = \dfrac{|\frac{1}{2}\pi - 2|}{1} = 2 - \frac{1}{2}\pi$.

Hence convergence of the Newton-Raphson process is guaranteed for any starting value x_0 such that $\frac{1}{2}\pi \leqslant x_0 \leqslant 2$. The results obtained with $x_0 = 2$ are shown on page 115.

7.7.3. *Practical approach*

It is not always easy to find theoretically an interval surrounding a root and such that the Newton-Raphson process will converge to this root for all starting values in the interval. For this reason the following approach will often be found to be useful. We have already seen that the Newton-Raphson process for *any* distinct root is second order so that $|F'(\lambda)| = 0$ and so certainly $|F'(\lambda)| < 1$. Hence, for all the functions F with which we will be concerned, there will be an interval I, surrounding λ, such that $|F'(x)| < 1$ for all x in I. Therefore, for *any* distinct root, the Newton-Raphson formula will give a convergent process when used with a starting value sufficiently close to λ, that is, in the interval I. Hence the convergence of the process is seen to depend solely on the choice of starting value. (Note the difference between this situation and that which arose when considering first-order methods. There a given formula might never give a convergent process for a particular root, no matter what starting value was used.) Therefore the only problem, using the Newton-Raphson process for a distinct root, is to obtain a starting value sufficiently close to the required root. This may be rather difficult. The procedure adopted is then

   1. Obtain an interval $[a, b]$ containing the required root.

   2. Choose a starting value x_0 in this interval. A suitable value might be obtained by applying the method of false position in the given interval; that is, use the value

$$x_0 = \frac{bf(a) - af(b)}{f(a) - f(b)}.$$

3. Start the iterations. Calculate $x_1, x_2, \ldots$.

4. If the process is not converging, begin again with a more accurate starting value x_0.

7.7.4. Order of the method

The Newton-Raphson method for a root λ of the equation $f(x) = 0$ is given by $x_{n+1} = F(x_n)$ where $F(x) = x - \dfrac{f(x)}{f'(x)}$.

Then
$$F'(x) = \frac{f(x)f''(x)}{[f'(x)]^2}$$

and so $F'(\lambda) = 0$ (since $f(\lambda) = 0$) unless $f'(\lambda) = 0$.

Then
$$F''(x) = \frac{(f'(x))^2 f''(x) + f(x)f'(x)f'''(x) - 2f(x)(f''(x))^2}{(f'(x))^3}$$

and so
$$F''(\lambda) = \frac{f''(\lambda)}{f'(\lambda)} \neq 0 \text{ in general.}$$

Thus the Newton-Raphson method is generally a second-order method. Difficulties arise if $f'(\lambda) = 0$, that is, if λ is a multiple root of the equation $f(x) = 0$. This case will be discussed later.

7.7.5. Particular applications of the Newton-Raphson method

We can find the square root of a positive constant a by using the Newton-Raphson method. The problem is to obtain the positive root of the equation $x^2 - a = 0$ or $f(x) = 0$ with $f(x) = x^2 - a$.

Then the Newton-Raphson method in this case is given by

$$x_{n+1} = x_n - \frac{x_n^2 - a}{2x_n}$$

$$= \frac{1}{2}\left(x_n + \frac{a}{x_n}\right).$$

It can be shown, using the conditions of 7.7.2, that this method will converge to $\sqrt{a}$ for any positive starting value x_0.

Now for the Newton-Raphson method we have

$$e_{n+1} \simeq \tfrac{1}{2}e_n^2 F_n''(\lambda) = \tfrac{1}{2}e_n^2 \frac{f''(\lambda)}{f'(\lambda)}.$$

But $f(x) = x^2 - a$ and so

$$e_{n+1} \simeq \tfrac{1}{2}e_n^2 \left(\frac{2}{2\lambda}\right) = \frac{e_n^2}{2\lambda}.$$

Thus, since $\lambda > 0$, all the e_i for $i = 1, 2, \ldots$ will be > 0. e_0 depends on the choice of starting value x_0. Hence, no matter whether x_0 is chosen less than or greater than λ, all the subsequent iterates will be *greater* than λ, that is, all the iterates $x_1, x_2, \ldots$ will be overestimates of the root λ.

Example. Determine $\sqrt{30}$ correct to 16 decimal places.

We can start with any positive value for x_0. However, since it is easily seen that $\sqrt{30}$ lies somewhere between 5 and 6, we shall take $x_0 = 5.5$. Then $|e_0| < 0.5$.

Now for the Newton-Raphson method we have

$$e_{n+1} \simeq \tfrac{1}{2}e_n^2 F''(\lambda) = \tfrac{1}{2}e_n^2 \frac{f''(\lambda)}{f'(\lambda)}.$$

In this case, $f(x) = x^2 - 30$ and so

$$e_{n+1} \simeq \frac{e_n^2}{2\lambda}.$$

But here $\lambda > 5$ and so $\dfrac{e_n^2}{2\lambda} < 0.1 e_n^2$.

Therefore an estimate of the error e_1 in the first iterate x_1 is

$0.1(0.5)^2 \simeq 0.03$ (rounding to one significant figure).

Hence we shall determine x_1 to two decimal places.

Then an estimate of the error e_2 in the second iterate x_2 is

$0.1(0.03)^2 \simeq 0.0001$ (rounding to one significant figure),

and we shall calculate x_2 to four decimal places.

Then an estimate of the error e_3 in the third iterate x_3 is

$0.1(0.0001)^2 = 0.000\,000\,001$

and we shall calculate x_3 to nine decimal places.

Then an estimate of the error e_4 in the fourth iterate x_4 is

$0.1(0.000\,000\,001)^2 = 0.000\,000\,000\,000\,000\,000\,1.$

E

Hence, adopting the above procedure for x_1, x_2, x_3, the fourth iterate x_4 will certainly give $\sqrt{30}$ correct to 16 decimal places.

$$x_1 = \tfrac{1}{2}\left(x_0 + \frac{30}{x_0}\right)$$

= 5·48 to two decimal places.

$$x_2 = \tfrac{1}{2}\left(x_1 + \frac{30}{x_1}\right)$$

= 5·4772 to four decimal places.

$$x_3 = \tfrac{1}{2}\left(x_2 + \frac{30}{x_2}\right)$$

= 5·477 225 575 to nine decimal places.

$$x_4 = \tfrac{1}{2}\left(x_3 + \frac{30}{x_3}\right)$$

= 5·477 225 575 255 166 1 to 16 decimal places.

Hence to 16 decimal places $\sqrt{30}$ is 5·477 225 575 255 166 1.

Note that, if tables of square roots are available, then the value for $\sqrt{a}$ obtained from them may be used as x_0. This will probably result in fewer iterations being required to obtain a given accuracy. For example, from tables $\sqrt{30} \simeq 5·477226$ and so, taking x_0 equal to this value, only two iterations are required to give the value of $\sqrt{30}$ correct to 16 decimal places. (The reader should verify this for himself.)

Similar processes can be derived to obtain the third, fifth, sixth roots, etc., of any given constant a.

To determine the pth root of a we must solve $x^p = a$ or $x^p - a = 0$. The Newton-Raphson method then gives

$$x_{n+1} = x_n - \frac{x_n^p - a}{px_n^{p-1}} = \frac{a + (p-1)x_n^p}{px_n^{p-1}}.$$

Note that if p is even then a must be positive and x_0 must be chosen positive. If p is odd and a is negative, x_0 should be chosen negative.

Example. Determine $\sqrt[3]{7}$ correct to four decimal places.
The Newton-Raphson method is given by

$$x_{n+1} = \frac{7 + 2x_n^3}{3x_n^2}.$$

With $x_0 = 2$ we obtain

$$x_1 = \frac{7 + 2(2)^3}{3(2)^2} = 1 \cdot 9 \text{ (rounding to one decimal place)}.$$

$$x_2 = \frac{7 + 2(1 \cdot 9)^3}{3(1 \cdot 9)^2} = 1 \cdot 913 \text{ (rounding to three decimal places)}.$$

$$x_3 = \frac{7 + 2(1 \cdot 913)^3}{3(1 \cdot 913)^2} = 1 \cdot 9129 \text{ (rounding to four decimal places)}.$$

Consideration of the error estimates, as in the last example, indicates that there is no need to continue the process any further. This could, of course, be confirmed by carrying out one further step. Hence $\sqrt[3]{7}$ is $1 \cdot 9129$ correct to four decimal places.

7.7.6. Flow diagram

The flow diagram for the Newton-Raphson process is the same as that for simple iteration (see § 7.3.1). The function F will of course be different. For the Newton-Raphson process for the equation $f(x) = 0$,

$$F(x) = x - \frac{f(x)}{f'(x)}.$$

EXERCISES

1. Use the Newton-Raphson formula with starting value 0 to obtain, correct to three decimal places, the root of $x^3 + x - \frac{1}{2} = 0$.

2. Use the Newton-Raphson formula with starting value 1 to obtain, correct to two decimal places, the root of $\sin x + x - 1 = 0$.

3. Use the Newton-Raphson formula to obtain, correct to four significant figures, the root of the equation $\sin x - x + \frac{1}{2} = 0$.

4. Find, correct to four significant figures, the only positive solution of the equation $x^3 + 2x - 6 = 0$ by elementary iteration and by the Newton-Raphson method.

5. Show that the equation $x = e^{-x}$ has a root in the interval $[\frac{1}{2}, 1]$ and use the Newton-Raphson formula to determine this root correct to three decimal places. (The derivative of the function e^{-x} is $-e^{-x}$.)

6. Use the Newton-Raphson formula to obtain $\sqrt{6}$ and $\sqrt{217}$, each to six significant figures, and check your results against standard tables.

7. Use the Newton-Raphson formula to devise an iteration procedure for the calculation of the reciprocal of a number without using the process of division. Hence evaluate $\frac{1}{17}$ to six significant figures and check your result with standard tables.

8. Show that the equation $x = 3 \ln x$ has a root between 1·5 and 2 and another between 4·5 and 5.

Evaluate these roots correct to three decimal places using (i) elementary iteration, (ii) the Newton-Raphson method.

9. Show that the equation $x^2 = \sin \pi x$ has a root between $x = \frac{3}{4}$ and $x = \frac{5}{6}$ and that the iterative formula $x_{n+1} = \frac{1}{\pi} \sin^{-1}(x_n^2)$ guarantees convergence with any starting value x_0 in $\left[\frac{3}{4}, \frac{5}{6}\right]$ whereas the formula $x_{n+1} = (\sin \pi x_n)^{1/2}$ does not.

Determine the root correct to three decimal places using (i) the above convergent process and Aitken's δ^2-process and (ii) the Newton-Raphson process.

7.8. The solution of polynomial equations—the Birge-Vieta process

In this section we develop a systematic method of applying the Newton-Raphson process to determine all the real roots of polynomial equations.

Let $p(x)$ denote the mth degree polynomial

$$a_m x^m + a_{m-1} x^{m-1} + \cdots + a_2 x^2 + a_1 x + a_0 \quad (a_m \neq 0).$$

Using the Newton-Raphson process to determine a root of $p(x) = 0$ we have

$$x_{n+1} = x_n - \frac{p(x_n)}{p'(x_n)}.$$

But $p(x_n)$ can of course be evaluated by the nesting method described in §3.1 by generating the sequence $\{b_r\}, r = m, m-1, \ldots, 1, 0$, where

$$b_m = a_m, \quad b_{m-k} = x_n b_{m-k+1} + a_{m-k} \quad \text{for} \quad k = 1, 2, \ldots, m.$$

Then $p(x_n) = b_0(x_n)$.

But, since $p(x)$ may be expressed in the form

$$p(x) = (x - x_n)q(x) + b_0$$

we see that $$p'(x) = q(x) + (x - x_n)q'(x)$$

and so $$p'(x_n) = q(x_n).$$

However q is a polynomial of degree $(n-1)$ and so can be evaluated at x_n by generating the sequence $\{c_r\}$ where

$$c_m = b_m, \quad c_{m-k} = x_n c_{m-k+1} + b_{m-k} \quad \text{for} \quad k = 1, 2, \ldots, m-1.$$

Now in practice we will only be able to obtain approximate values for the coefficients of the polynomial $q(x)$ unless λ is known exactly. Thus we will only obtain a polynomial which approximates the reduced polynomial $q(x)$. Hence, even if this polynomial could be solved exactly, we could only hope to obtain approximations to the remaining roots of the given polynomial $p(x)$. If the process is being repeated several times to obtain a number of the roots of the given polynomial, then errors of the above sort can build up. It is found that to obtain greatest accuracy the roots should be determined in increasing order of magnitude.

Each step in the Birge-Vieta process may be conveniently recorded in a similar way to that used in the synthetic division process.

For example, consider one step in the process to find a root of the equation $2x^3 + 7x^2 - 7x - 11 = 0$ starting with the approximation $x_0 = 1.5$.

Let $p(x) \equiv 2x^3 + 7x^2 - 7x - 11$.

Then
$$x_1 = x_0 - \frac{p(x_0)}{p'(x_0)} = 1.5 - \frac{p(1.5)}{p'(1.5)}.$$

We evaluate $p(1.5)$ by the synthetic division process

		2	7	-7	-11
1.5			3	15	12
		2	10	8	1

Hence $p(1.5) = 1$.

Then we evaluate $p'(1.5)$ by carrying out the synthetic division process on the polynomial $2x^2 + 10x + 8$.

		2	10	8
1.5			3	19.5
		2	13	27.5

Hence $p'(1.5) = 27.5$.

Therefore
$$x_1 = 1.5 - \frac{1}{27.5} \simeq 1.464.$$

The middle lines of each of the above two tables can be omitted and the remainders joined to give one table.

		2	7	-7	-11
1.5		2	10	8	1
1.5		2	13	27.5	

For the general polynomial of mth degree $a_m x^m + a_{m-1} x^{m-1} + \cdots + a_1 x + a_0$ the step of the iterative process which obtains x_{n+1} from x_n would be recorded as shown:

	a_m	a_{m-1}	a_{m-2}	$\cdots$	a_2	a_1	a_0
x_n	b_m	b_{m-1}	b_{m-2}	$\cdots$	b_2	b_1	$\underline{b_0}$
x_n	c_m	c_{m-1}	c_{m-2}	$\cdots$	c_2	$\underline{c_1}$	

Then
$$x_{n+1} = x_n - \frac{b_0}{c_1}.$$

The b_r are calculated first using

$$b_m = a_m, \quad b_{m-k} = x_n b_{m-k+1} + a_{m-k} \quad \text{for} \quad k = 1, 2, \ldots, m,$$

then the c_r are calculated using

$$c_m = b_m, \quad c_{m-k} = x_n c_{m-k+1} + b_{m-k} \quad \text{for} \quad k = 1, 2, \ldots, m-1.$$

Example. Determine the two real roots of the polynomial $x^4 - 6x^3 - 2x^2 + 58x - 50$ correct to three decimal places, given that there is a root near each of 1 and -3.

Since the roots should be calculated in increasing order of magnitude we shall attempt to obtain the root near $x = 1$ first. Thus take $x_0 = 1$. The calculation then proceeds as follows:

	1	-6	-2	58	-50
1	1	-5	-7	51	$\underline{1}$
1	1	-4	-11	$\underline{40}$	

Hence
$$x_1 = 1 - \frac{1}{40} = 0.975.$$

	1	-6	-2	58	-50
0.975	1	-5.0250	-6.8994	51.2731	$\underline{-0.0087}$
0.975	1	-4.0500	-10.8481	$\underline{40.6962}$	

Hence
$$x_2 = 0.975 + \frac{0.0087}{40.6962} \simeq 0.9752.$$

Thus there is a root 0.975 correct to three decimal places. For the root near -3 we consider the reduced polynomial

$$x^3 - 5.0250x^2 - 6.8994x + 51.2731$$

and take $x_0 = -3$.

The calculation then proceeds as follows:

	1	−5·0250	−6·8994	51·2731
−3	1	−8·0250	17·1756	−0·2537
−3	1	−11·0250	50·2506	

Hence $x_1 = -3 + \dfrac{0·2537}{50·2506} \simeq -2·995.$

Now we are very close to the required root, and so we shall carry out the next (and final) iteration using the original polynomial. This is advisable since the coefficients we have been using for the reduced polynomial are only approximate.

We then obtain the following results:

	1	−6	−2	58	−50
−2·995	1	−8·9950	24·9400	−16·6953	0·0024
−2·995	1	−11·9900	60·8500	−198·9410	

Hence $x_2 = -2·995 + \dfrac{0·0024}{198·9410} = -2·995$ correct to three decimal places.

Thus there is a root −2·995 correct to three decimal places.

The remaining two roots of the given quartic are now approximately equal to the roots of the quadratic equation

$$x^2 - 8·025x + 17·176 = 0.$$

The coefficients of this quadratic are, however, inexact and better approximations to them can be obtained by dividing the original polynomial by the factors $(x - 0·975)$ and $(x + 2·995)$ as follows:

	1	−6	−2	58	−50
0·975	1	−5·025	−6·899	51·273	−0·009
−2·995	1	−8·020	17·121		

Thus we obtain the quadratic equation

$$x^2 - 8·020x + 17·121 = 0$$

which can now be solved to give the remaining two (complex in this case) roots of the original quartic.

EXERCISES

1. The equation $x^3 - 10x + 4 = 0$ has one root in each of the intervals $[-4, -3]$, $[0, 1]$ and $[2, 3]$. Use the Birge-Vieta process to determine these roots correct to three significant figures.

2. Show that the equation $x^3 - 5x^2 + 5x + 1 = 0$ has one root between -1 and 0, another between 1 and 2, and the third between 3 and 4. Use the Birge-Vieta process to obtain all the roots correct to three decimal places.

3. Obtain, correct to three decimal places, the only two real roots of the polynomial

$$x^4 + 2x^3 + 7x^2 - 11 = 0.$$

7.9. Equations with nearly equal roots

We now consider an equation of the form $f(x) = 0$ which has two nearly equal roots near $x = \lambda$. In practice this will include the case of a double root at $x = \lambda$.

Then $\qquad\qquad f(\lambda) \simeq 0 \quad \text{and} \quad f'(\lambda) \simeq 0.$

[For a double root at $x = \lambda$ we will have $f(x) = (x - \lambda)^2 q(x)$ and so $f'(x) = 2(x - \lambda)q(x) + (x - \lambda)^2 q'(x)$. Hence $f(\lambda) = 0$ and $f'(\lambda) = 0$.]

We see that for values of x close to the root λ, $f'(x)$ will become very small, and so the Newton-Raphson method will get into difficulty. This difficulty can be avoided by looking for a root of $f'(x) = 0$ in a neighbourhood of λ. If $f(x) = 0$ has only two nearly equal roots near $x = \lambda$, then $f'(x) = 0$ will have only a single root near $x = \lambda$, and so the above difficulty of applying the Newton-Raphson method should not exist. From the approximation to the root of $f'(x) = 0$ we can then obtain approximations to each of the two corresponding roots of $f(x) = 0$ as follows.

Let μ be a root of $f'(x) = 0$ such that $\mu + \varepsilon$ (where ε is small) is a root of $f(x) = 0$.

Then $\qquad\qquad\qquad f(\mu + \varepsilon) = 0.$

But, using Taylor's series expansion,* we obtain

$$f(\mu + \varepsilon) = f(\mu) + \varepsilon f'(\mu) + \tfrac{1}{2}\varepsilon^2 f''(\mu) + \dots$$
$$= f(\mu) + \tfrac{1}{2}\varepsilon^2 f''(\mu) +$$

since $f'(\mu) = 0$.

Hence, since ε is small,

$$0 \simeq f(\mu) + \tfrac{1}{2}\varepsilon^2 f''(\mu)$$

and so $\qquad\qquad\qquad \varepsilon^2 \simeq -\dfrac{2f(\mu)}{f''(\mu)}.$

---

* See Appendix 1.

Now if $f(\mu)$ and $f''(\mu)$ are of opposite sign, that is, if f is positive and has a maximum at $x = \mu$ (Figure 18a) or f is negative and has a minimum at $x = \mu$ (Figure 18b) then

$$-\frac{2f(\mu)}{f''(\mu)} > 0.$$

Hence ε is real and the values $(\mu + \varepsilon)$ and $(\mu - \varepsilon)$ may be used as first approximations for the two roots of $f(x) = 0$.

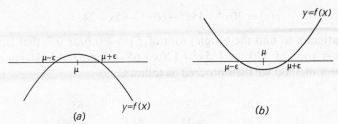

Figure 18

If $f(\mu)$ and $f''(\mu)$ have the same sign, that is, if f is negative and has a maximum at $x = \mu$ (Figure 19a) or f is positive and has a minimum at $x = \mu$ (Figure 19b) then

$$-\frac{2f(\mu)}{f''(\mu)} < 0.$$

Hence ε is not a real number and the given equation $f(x) = 0$ does not have real roots near $x = \mu$.

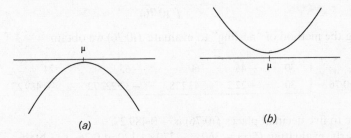

Figure 19

Equations such as we are considering here are said to be *ill-conditioned**
since relatively small changes in the constants in the function f can

---

* For a discussion of ill-conditioning in this context see Appendix 4.

produce quite large changes in the roots. Conversely too, a relatively large change in x may have little effect on the value of the function f.

For example, the quadratic equation $x^2 - 1\cdot498x + 0\cdot561 = 0$ has exact roots $0\cdot75$ and $0\cdot748$, while the quadratic equation $x^2 - 1\cdot500x + 0\cdot561 = 0$ has roots $0\cdot789$ and $0\cdot711$ correct to three decimal places. Also, evaluating the quadratic $x^2 - 1\cdot498x + 0\cdot561$ at $0\cdot749$ gives $-0\cdot000\,001$.

Example. Determine, correct to three decimal places, the two roots of the equation $30x^4 - 45x^3 + 60x^2 - 65x + 24 = 0$ which are near $0\cdot7$.

If $$f(x) = 30x^4 - 45x^3 + 60x^2 - 65x + 24$$

we first attempt to find the (single) root of $f'(x) = 0$ near $0\cdot7$, that is, we look for a root of $120x^3 - 135x^2 + 120x - 65 = 0$ near $0\cdot7$. Using the Birge-Vieta method we then proceed as follows.

	120	-135	120	-65
$0\cdot7$	120	-51	$84\cdot3$	$-5\cdot99$
$0\cdot7$	120	33	$107\cdot4$	

Therefore $$x_1 = 0\cdot7 + \frac{5\cdot99}{107\cdot4}$$

$$\simeq 0\cdot76 \text{ (rounding to two decimal places).}$$

Then, using this approximation to the root of $f'(x) = 0$, we have, in the above notation,

$$\varepsilon^2 = -\frac{2f(0\cdot76)}{f''(0\cdot76)}.$$

Using the method of "nesting" to evaluate $f(0\cdot76)$ we obtain

	30	-45	60	-65	24
$0\cdot76$	30	$-22\cdot2$	$43\cdot128$	$-32\cdot222\,72$	$-0\cdot489\,27$

Hence to five decimal places $f(0\cdot76)$ is $-0\cdot489\,27$.
Similarly, evaluating $f''(x) \equiv 360x^2 - 270x + 120$ at $0\cdot76$, we obtain

	360	-270	120
$0\cdot76$	360	$3\cdot6$	$122\cdot736$

Hence $$f''(0\cdot76) = 122\cdot736 \text{ (exactly).}$$

Therefore $\quad \varepsilon^2 = \dfrac{2(0.489\,27)}{122.736}$

$\simeq 0.0080$ (rounding to two significant figures).

Hence $\qquad \varepsilon \simeq \pm 0.09$ (rounding to one significant figure).

Therefore we should use 0·85 and 0·67 as starting approximations for determining the two roots of the given equation

$$30x^4 - 45x^3 + 60x^2 - 65x + 24 = 0.$$

We shall determine the smaller of the two roots first.
Using the Birge-Vieta method the calculations then proceed as follows:

	30	−45	60	−65	24
0·67	30	−24·9	43·317	−35·977 61	−0·105 00
0·67	30	−4·8	40·101	−9·109 94	

Therefore $\quad x_1 = 0.67 - \dfrac{0.105\,00}{9.109\,94}$

$\simeq 0.658$ (rounding to three decimal places).

	30	−45	60	−65	24
0·658	30	−25·26	43·378 92	−36·456 67	0·011 51
0·658	30	−5·52	39·746 76	−10·303 30	

Therefore $\quad x_2 = 0.658 + \dfrac{0.011\,51}{10.303\,30}$

$\simeq 0.6591$ (rounding to four decimal places).

	30	−45	60	−65	24
0·6591	30	−25·227	43·372 88	−36·412 93	0·000 24
0·6591	30	−5·454	39·778 15	−10·195 15	

Therefore $\quad x_4 = 0.6591 + \dfrac{0.000\,24}{10.195\,15}$

$\simeq 0.6591$ (rounding to four decimal places).

Hence one root, correct to three decimal places, is 0·659.

To determine the second root we continue the process with the reduced polynomial $30x^3 - 25·227x^2 + 43·372\,88x - 36·412\,93$ and $x_0 = 0·85$.

	30	−25·227	43·372 88	−36·412 93
0·85	30	0·273	43·604 93	0·651 26
0·85	30	25·773	65·511 98	

Therefore $x_1 = 0·85 - \dfrac{0·651\,26}{65·511\,98}$

$\simeq 0·8401$ (rounding to four decimal places).

The next step is carried out using the original polynomial.

	30	−45	60	−65	24
0·8401	30	−19·797	43·368 54	−28·566 09	−0·001 63
0·8401	30	5·406	47·910 12	11·683 20	

Therefore $x_2 = 0·8401 + \dfrac{0·001\,63}{11·683\,20}$

$\simeq 0·8400$ (rounding to four decimal places).

Hence the second root, correct to three decimal places, is 0·840.

EXERCISES

1. Find, correct to three decimal places, the positive roots of the equation
$x^3 - 12x + 15 = 0$.
[It is easily seen that there is a root between $x = 1$ and $x = 2$ and another root between $x = 2$ and $x = 3$. Also, $f'(x)$ becomes extremely small as x becomes close to 2. The problem should be treated as if there are two roots near $x = 2$.]

2. Given that the equation
$$x^4 - 1{\cdot}5x^3 + 4{\cdot}56x^2 - 6x + 2{\cdot}25 = 0$$

has two roots near $x = 0{\cdot}75$, obtain these roots correct to three decimal places, assuming that the coefficients in the given equation are exact.

The solution of systems of linear equations

8

THE PROBLEM WE ARE CONCERNED WITH IN THIS CHAPTER IS THAT OF solving systems of linear equations such as the four by four system

$$3x_1 - x_2 + 7x_3 + 4x_4 = 1$$
$$2x_1 + 3x_2 + 5x_3 - 2x_4 = 3$$
$$x_1 + 2x_2 - 3x_3 + 6x_4 = 5$$
$$-5x_1 + 4x_2 + x_3 - 2x_4 = 2.$$

The general $n \times n$ system has the form

$$a_{11}x_1 + a_{12}x_2 + \cdots + a_{1n}x_n = b_1$$
$$a_{21}x_1 + a_{22}x_2 + \cdots + a_{2n}x_n = b_2$$
$$\cdots\cdots\cdots\cdots\cdots\cdots\cdots\cdots\cdots\cdots\cdots\cdots$$
$$a_{n1}x_1 + a_{n2}x_2 + \cdots + a_{nn}x_n = b_n$$

that is, n linear equations in n unknowns. We shall only consider systems for which a unique solution exists. In matrix notation then we have $Ax = b$ where

$$A = \begin{pmatrix} a_{11} & a_{12} & \cdots & a_{1n} \\ a_{21} & a_{22} & \cdots & a_{2n} \\ \cdots\cdots\cdots\cdots\cdots\cdots\cdots \\ a_{n1} & a_{n2} & \cdots & a_{nn} \end{pmatrix}$$

is a non-singular $n \times n$ matrix,*

$$b = \begin{pmatrix} b_1 \\ b_2 \\ \vdots \\ b_n \end{pmatrix}$$

---

* A non-singular matrix is one whose determinant is not zero.

is a given n-dimensional vector and

$$x = \begin{pmatrix} x_1 \\ x_2 \\ \vdots \\ x_n \end{pmatrix}$$

is an n-dimensional vector whose components it is desired to find.

Problems of this type occur naturally in many branches of applied mathematics. Also, when solving more complicated problems, a first step is often to reduce the problem to one of the above type.

There are two main classes of methods for solving linear systems of equations numerically, namely *direct* and *iterative*. Iterative methods are essentially an extension of the methods of the last chapter but, whereas there we carried out the iterative processes for a scalar, it is now a vector which is required. These methods, which are normally applied only when the matrix of the system is sparse (that is has many zero elements), are beyond the scope of this book. The remainder of this chapter will be concerned with direct methods, that is, methods which go directly to the solution in "one step". This is in contrast to iterative methods which repeat a process over and over again until convergence (to within a specified tolerance) is obtained. Direct methods have a fixed predictable number of operations (that is, additions, subtractions, multiplications and divisions) but are affected by inaccuracies due to round-off errors.

8.1. Gaussian elimination

The method of Gaussian elimination which we shall describe first reduces any given system to a "triangular" system of equations which is solved by the method of back substitution. Firstly we illustrate back substitution by solving the following system of equations:

$$\begin{aligned} 3x_1 - 2x_2 - 4x_3 &= 3 \\ 4x_2 + 3x_3 &= 2 \\ 2x_3 &= 4. \end{aligned}$$

From the last equation we obtain $x_3 = 2$, and then from the second we obtain $4x_2 = 2 - 3x_3 = -4$, and so $x_2 = -1$. Finally from the first equation we obtain $3x_1 = 3 + 2x_2 + 4x_3 = 9$ and so $x_1 = 3$. We now illustrate the complete process of Gaussian elimination in the following example.

Example. Solve the system

$$4x_1 + 2x_2 + 3x_3 = 6$$
$$5x_1 - 5x_2 + 3x_3 = 3$$
$$3x_1 - x_2 - 2x_3 = 2.$$

If we leave the first equation unaltered, then subtract $\frac{5}{4}$ times this equation from the second to obtain a new second equation, and $\frac{3}{4}$ times the first from the third to obtain a new third equation, we obtain the system

$$4x_1 \; + 2x_2 \; + 3x_3 = 6$$
$$-7 \cdot 5x_2 - 0 \cdot 75x_3 = -4 \cdot 5$$
$$-2 \cdot 5x_2 - 4 \cdot 25x_3 = -2 \cdot 5$$

Now if we leave the first two equations in this system unaltered and then subtract $(\frac{2 \cdot 5}{7 \cdot 5})$ times the second equation from the third equation to obtain a new third equation, we obtain the system

$$4x_1 \; + 2x_2 \; + 3x_3 = 6$$
$$-7 \cdot 5x_2 - 0 \cdot 75x_3 = -4 \cdot 5$$
$$-4x_3 = -1$$

This is a "triangular" system and so is easily solved, as above, to give

$$x_3 = 0 \cdot 25$$
$$x_2 = -\tfrac{1}{7 \cdot 5}(-4 \cdot 5 + 0 \cdot 75x_3) = 0 \cdot 575$$

and
$$x_1 = \tfrac{1}{4}(6 - 2x_2 - 3x_3) = 1 \cdot 025.$$

This illustrates the process of Gaussian elimination. The working can be conveniently and economically recorded in the following tabular form:

left-hand coefficients			right-hand coefficient
a_{i1}	a_{i2}	a_{i3}	b_i
4	2	3	6
5	−5	3	3
3	−1	−2	2
	−7·5	−0·75	−4·5
	−2·5	−4·25	−2·5
		−4	−1
Roots 1·025	0·575	0·25	

Note that at each stage we do not write down again the coefficients of any equation which has remained unaltered from the previous stage.

Now consider the general 3×3 system:

$$a_{11}x_1 + a_{12}x_2 + a_{13}x_3 = b_1$$

$$a_{21}x_1 + a_{22}x_2 + a_{23}x_3 = b_2$$

$$a_{31}x_1 + a_{32}x_2 + a_{33}x_3 = b_3$$

in which $a_{11} \neq 0$. (Note that since the equations have a unique solution we can always arrange that $a_{11} \neq 0$ by rearranging the given equations if necessary.) Eliminate x_1 from the second equation by subtracting $m_{21} = a_{21}/a_{11}$ times the first equation from the second equation. Similarly, eliminate x_1 from the third equation by subtracting $m_{31} = a_{31}/a_{11}$ times the first equation from it. Thus we obtain the system

$$a_{11}x_1 + a_{12}x_2 + a_{13}x_3 = b_1$$

$$a_{22}^{(1)}x_2 + a_{23}^{(1)}x_3 = b_2^{(1)}$$

$$a_{32}^{(1)}x_2 + a_{33}^{(1)}x_3 = b_3^{(1)}$$

in which

$$a_{22}^{(1)} = a_{22} - m_{21}a_{12}, \quad a_{23}^{(1)} = a_{23} - m_{21}a_{13}, \quad b_2^{(1)} = b_2 - m_{21}b_1,$$

$$a_{32}^{(1)} = a_{32} - m_{31}a_{12}, \quad a_{33}^{(1)} = a_{33} - m_{31}a_{13}, \quad b_3^{(1)} = b_3 - m_{31}b_1.$$

Now eliminate x_2 from the third equation in the above system by subtracting $m_{32} = a_{32}^{(1)}/a_{22}^{(1)}$ times the second equation from it. (Note that if $a_{22}^{(1)} = 0$ then it will be necessary to interchange the second and third equations before carrying out this step.) Thus we obtain the following triangular system,

$$a_{11}x_1 + a_{12}x_2 + a_{13}x_3 = b_1$$

$$a_{22}^{(1)}x_2 + a_{23}^{(1)}x_3 = b_2^{(1)}$$

$$a_{33}^{(2)}x_3 = b_3^{(2)}$$

in which $a_{33}^{(2)} = a_{33}^{(1)} - m_{32}a_{23}^{(1)}$ and $b_3^{(2)} = b_3^{(1)} - m_{32}b_2^{(1)}$.

Hence $x_3 = b_3^{(2)}/a_{33}^{(2)}$ and then, from the second equation of the above system, we can obtain x_2. From the first equation of the above system we can obtain x_1. This latter process is the process of back substitution and the complete algorithm is known as *Gaussian elimination*. The equations of the final (triangular system are called *pivotal equations* and the coefficients a_{11}, $a_{22}^{(1)}$, $a_{33}^{(2)}$ are called the *pivots* for the process.

In tabular form we have

m_{ij}	left-hand coefficients			right-hand coefficient
	a_{i1}	a_{i2}	a_{i3}	b_i
1	a_{11}	a_{12}	a_{13}	b_1
a_{21}/a_{11}	a_{21}	a_{22}	a_{23}	b_2
a_{31}/a_{11}	a_{31}	a_{32}	a_{33}	b_3
1		$a_{22}^{(1)}$	$a_{23}^{(1)}$	$b_2^{(1)}$
$a_{32}^{(1)}/a_{22}^{(1)}$		$a_{32}^{(1)}$	$a_{33}^{(1)}$	$b_3^{(1)}$
1			$a_{33}^{(2)}$	$b_3^{(2)}$
Roots from back substitution	x_1	x_2	$x_3 = b_3^{(2)}/a_{33}^{(2)}$	

Note that we have put $m_{11} = m_{22} = m_{33} = 1$.

The process which we have described above for a three by three system is, of course, easily generalized to the case of higher-order systems. For the $n \times n$ system

$$a_{11}x_1 + a_{12}x_2 + \cdots + a_{1n}x_n = b_1$$
$$a_{21}x_1 + a_{22}x_2 + \cdots + a_{2n}x_n = b_2$$
$$\dots\dots\dots\dots\dots\dots\dots\dots\dots\dots\dots\dots$$
$$a_{n1}x_1 + a_{n2}x_2 + \cdots + a_{nn}x_n = b_n$$

we proceed as follows:

Eliminate x_1 from the second equation by subtracting $m_{21} = a_{21}/a_{11}$ times the first equation from it. (The given equations should first be rearranged, if necessary, to ensure that $a_{11} \neq 0$.) Similarly eliminate x_1 from each of the following equations by subtracting $m_{i1} = a_{i1}/a_{11}$ times the first equation from the ith equation for $i = 2, 3, \ldots, n$. Thus we obtain the system

$$a_{11}x_1 + a_{12}x_2 + \cdots + a_{1n}x_n = b_1$$
$$a_{22}^{(1)}x_2 + \cdots + a_{2n}^{(1)}x_n = b_2^{(1)}$$
$$a_{32}^{(1)}x_2 + \cdots + a_{3n}^{(1)}x_n = b_3^{(1)}$$
$$\cdots\cdots\cdots\cdots\cdots\cdots\cdots\cdots\cdots$$
$$a_{n2}^{(1)}x_2 + \cdots + a_{nn}^{(1)}x_n = b_n^{(1)}$$

where $a_{ij}^{(1)} = a_{ij} - m_{i1}a_{1j}$ and $b_i^{(1)} = b_i - m_{i1}b$, for $i = 2, 3, \ldots, n$ and $j = 2, 3, \ldots, n$.

Now eliminate x_2 from each of the 3rd, 4th, $\ldots$, nth equations of this last system by subtracting $m_{i2} = a_{i2}^{(1)}/a_{22}^{(1)}$ times the second equation from the ith equation for $i = 3, 4, \ldots, n$. (The equations must again be re-arranged, if necessary, to ensure that $a_{22}^{(1)} \neq 0$ before carrying out this step.) Thus we obtain the system

$$a_{11}x_1 + a_{12}x_2 + \cdots + a_{1n}x_n = b_1$$
$$a_{22}^{(1)}x_2 + \cdots + a_{2n}^{(1)}x_n = b_2^{(1)}$$
$$a_{33}^{(2)}x_3 + \cdots + a_{3n}^{(2)}x_n = b_3^{(2)}$$
$$a_{43}^{(2)}x_3 + \cdots + a_{4n}^{(2)}x_n = b_4^{(2)}$$
$$\cdots\cdots\cdots\cdots\cdots\cdots\cdots\cdots\cdots$$
$$a_{n3}^{(2)}x_3 + \cdots + a_{nn}^{(2)}x_n = b_n^{(2)}$$

where $a_{ij}^{(2)} = a_{ij}^{(1)} - m_{i2}a_{2j}^{(1)}$, $b_i^{(2)} = b_i^{(1)} - m_{i2}b_2^{(1)}$ for $i = 3, 4, \ldots, n$ and $j = 3, 4, \ldots, n$.

This process is repeated until the following triangular system is obtained.

$$a_{11}x_1 + a_{12}x_2 + \cdots + a_{1n}x_n = b_1$$
$$a_{22}^{(1)}x_2 + \cdots + a_{2n}^{(1)}x_n = b_2^{(1)}$$
$$a_{33}^{(2)}x_3 + \cdots + a_{3n}^{(2)}x_n = b_3^{(2)}$$
$$\cdots\cdots\cdots\cdots\cdots\cdots\cdots\cdots\cdots$$
$$a_{nn}^{(n-1)}x_n = b_n^{(n-1)}.$$

Then $x_n = b_n^{(n-1)}/a_{nn}^{(n-1)}$ and hence $x_{n-1}, x_{n-2}, \ldots, x_2, x_1$ are obtained in that order by back substitution. The equations of this last system are the pivotal equations, and the coefficients are the pivots. The process can be summarized in tabular form as for the 3×3 case as follows:

Multipliers	left-hand coefficients		right-hand coefficient
m_{ij}	a_{i1}	$a_{i2} \cdots a_{in}$	b_i
1	a_{11}	$a_{12} \cdots a_{1n}$	b_1
a_{21}/a_{11}	a_{21}	$a_{22} \cdots a_{2n}$	b_2
$\vdots$	$\vdots$	$\vdots \qquad \vdots$	$\vdots$
a_{n1}/a_{11}	a_{n1}	$a_{n2} \cdots a_{nn}$	b_n
1		$a_{22}^{(1)} \cdots a_{2n}^{(1)}$	$b_2^{(1)}$
$a_{32}^{(1)}/a_{22}^{(1)}$		$a_{32}^{(1)} \cdots a_{3n}^{(1)}$	$b_3^{(1)}$
$\vdots$		$\vdots \qquad \vdots$	$\vdots$
$a_{n2}^{(1)}/a_{22}^{(1)}$		$a_{n2}^{(1)} \cdots a_{nn}^{(1)}$	$b_n^{(1)}$
$\vdots$		$\cdots \quad \vdots$	$\vdots$
1		$a_{nn}^{(n-1)}$	$b_n^{(n-1)}$
Roots from back substitution	α_1	$\alpha_2 \cdots \alpha_n$	

The solution is then $x_1 = \alpha_1$, $x_2 = \alpha_2, \ldots, x_n = \alpha_n$.

Example. Solve the following system of equations working to three decimal places throughout.

$$2 \cdot 801 x_1 - 1 \cdot 643 x_2 + 2 \cdot 017 x_3 = 3 \cdot 216$$
$$0 \cdot 762 x_1 + 1 \cdot 372 x_2 - 2 \cdot 631 x_3 = 1 \cdot 835$$
$$1 \cdot 314 x_1 - 4 \cdot 721 x_2 + 3 \cdot 596 x_3 = 1 \cdot 062.$$

Multipliers	left-hand coefficients			right-hand coefficient
m_{ij}	a_{i1}	a_{i2}	a_{i3}	b_i
1	2·801	−1·643	2·017	3·216
0·272	0·762	1·372	−2·631	1·835
0·469	1·314	−4·721	3·596	1·062
1		1·819	−3·180	0·960
−2·172		−3·950	2·650	−0·446
1			−4·257	1·639
Roots from back substitution	1·340	−0·145	−0·385	

EXERCISES

1. Use Gaussian elimination to solve approximately the following system of linear equations

$$2x_1 - x_2 - 4x_3 = 4$$
$$4x_1 + x_2 + 2x_3 = 4$$
$$3x_1 + 8x_2 - x_3 = 20.$$

Use three decimal places in the calculations.

2. Use Gaussian elimination to obtain an approximate solution of the following system of linear equations

$$3x_1 - 12x_2 + 8x_3 = 7$$
$$4x_1 + 7x_2 - 2x_3 = 4$$
$$7x_1 + 9x_2 + 5x_3 = 11.$$

Use three decimal places in the calculations.

3. Use Gaussian elimination to obtain an approximate solution of the following system of equations

$$7x_1 + 4x_2 + 17x_3 = -85$$
$$33x_1 + 16x_2 + 72x_3 = 359$$
$$24x_1 + 10x_2 + 57x_3 = 281.$$

Use three significant figures in the calculations.

4. Use Gaussian elimination to obtain an approximate solution of the following system of equations

$$0 \cdot 163x_1 + 1 \cdot 56x_2 - 8 \cdot 10x_3 = 2 \cdot 49$$
$$-0 \cdot 00154x_1 + 0 \cdot 796x_2 + 4 \cdot 69x_3 = 0 \cdot 652$$
$$6 \cdot 40x_1 - 0 \cdot 378x_2 + 0 \cdot 573x_3 = 0 \cdot 0328.$$

Use three significant figures in the calculations.

5. Construct a flow diagram for the back-substitution process.

6. Construct a flow diagram for the elimination process.

8.1.1. Checking procedures

In this section we shall describe two checking procedures which are useful when the calculations are being performed manually or using a desk calculator. These checking procedures are superfluous if a computer is being used, since, once a program has been developed and thoroughly tested, it should run without error (apart from perhaps a machine failure!) for different systems of equations.

The first check results in an extra column in the above tabular representation. We shall consider again the $n \times n$ system given in the previous section and change the notation a little by putting $b_i = a_{i\,n+1}$ for $i = 1, 2, \ldots, n$. The system of equations can now be written

$$a_{11}x_1 + a_{12}x_2 + \cdots + a_{1n}x_n = a_{1\,n+1}$$
$$a_{21}x_1 + a_{22}x_2 + \cdots + a_{2n}x_n = a_{2\,n+1}$$
$$\cdots\cdots\cdots\cdots\cdots\cdots\cdots\cdots\cdots\cdots\cdots$$
$$a_{n1}x_1 + a_{n2}x_2 + \cdots + a_{nn}x_n = a_{n\,n+1}.$$

Now let $\qquad a_{i\,n+2} = \displaystyle\sum_{j=1}^{n+1} a_{ij}$ for $i = 1, 2, \ldots, n;$

that is, $a_{i\,n+2}$ is the sum of all the coefficients in the ith equation or the sum of all the elements in the ith row of the table (excluding, of course, m_{ij}). The elements $a_{i\,n+2}$ for $i = 1, 2, \ldots, n$ are now written as an extra column on the right of the table and are then operated on to obtain $a_{i\,n+2}^{(1)}$, $a_{i\,n+2}^{(2)}$, etc., in exactly the same way as the elements a_{ij} for $j = 2, 3, \ldots, n+1$ are operated on to obtain $a_{ij}^{(1)}$, $a_{ij}^{(2)}$, etc., that is,

$$a_{i\,n+2}^{(1)} = a_{i\,n+2} - m_{i1}a_{1\,n+2}, \quad a_{i\,n+2}^{(2)} = a_{i\,n+2}^{(1)} - m_{i2}a_{2\,n+2}^{(1)}$$

etc., for appropriate values of i.

Therefore $\qquad a_{i\,n+2}^{(1)} = a_{i\,n+2} - m_{i1}a_{1\,n+2}$

$$= \sum_{j=1}^{n+1} a_{ij} - m_{i1} \sum_{j=1}^{n+1} a_{1j}$$

$$= \sum_{j=1}^{n+1} (a_{ij} - m_{i1}a_{1j}) = \sum_{j=1}^{n+1} a_{ij}^{(1)}.$$

Similarly $\qquad a_{i\,n+2}^{(2)} = \displaystyle\sum_{j=1}^{n+1} a_{ij}^{(2)}, \quad a_{i\,n+2}^{(3)} = \sum_{j=1}^{n+1} a_{ij}^{(3)}, \quad$ etc.

Thus, at each stage of the process, the elements in the $(n+2)$th column of coefficients should be equal, to within round-off errors, to the sum of the elements in the corresponding row (excluding of course the column of m_{ij}). If at any stage, due to round-off errors, the elements in the $(n+2)$th column obtained by the elimination algorithm is not exactly equal to the appropriate row sum, then it should be replaced by this (row sum) value for subsequent calculations. In this way the effectiveness of the check is not reduced by the accumulation of round-off errors. Thus this procedure checks each stage of the process separately. It provides a check on the elimination process. After we have described the other check we shall re-do the example on page 134 incorporating them both.

The second check, which we shall now describe, results in an extra row in the tabular representation.

Let
$$s_j = \sum_{i=1}^{n} a_{ij} \quad \text{for} \quad j = 1, 2, \ldots, n+2,$$

that is, the elements s_j are column sums of coefficients. These elements are written as an extra row along the bottom of the table. Now consider the scalar product of the vector $(s_1, s_2, \ldots, s_n)$ with the solution vector $(x_1, x_2, \ldots, x_n)$ obtained after back substitution.

This scalar product is equal to

$$s_1 x_1 + s_2 x_2 + \cdots + s_n x_n = \sum_{j=1}^{n} s_j x_j = \sum_{j=1}^{n} \left(\sum_{i=1}^{n} a_{ij} \right) x_j$$

$$= \sum_{i=1}^{n} \left(\sum_{j=1}^{n} a_{ij} x_j \right) = \sum_{i=1}^{n} a_{i\,n+1}$$

$$= s_{n+1}.$$

Thus, to within round-off error, the scalar product should be equal to s_{n+1}. Now the above result can also be written

$$\left(\sum_{i=1}^{n} a_{i1} \right) x_1 + \left(\sum_{i=1}^{n} a_{i2} \right) x_2 + \cdots + \left(\sum_{i=1}^{n} a_{in} \right) x_n = \sum_{i=1}^{n} b_i.$$

Hence we see that this check merely states that the solution obtained should satisfy, to within round-off error, the equation got by adding together all the equations of the given system. This check is a check on the back substitution. Since s_{n+2} has also been calculated we can perform the following additional check.

$$\sum_{j=1}^{n+1} s_j = \sum_{j=1}^{n+1} \sum_{i=1}^{n} a_{ij} = \sum_{i=1}^{n} \sum_{j=1}^{n+1} a_{ij} = \sum_{i=1}^{n} a_{i\,n+2} = s_{n+2}.$$

Thus the sum of the first $(n+1)$ of the s_j should be equal to s_{n+2}. This check should be exact and not subject to round-off error since s_{n+2} is just the sum of all the coefficients in the given system of equations obtained by summing the row sums, whereas $\sum_{j=1}^{n+1} s_j$ is the same quantity obtained by summing all the column sums.

We shall now re-do the example on page 134 incorporating the above checks.

Multipliers	left-hand coefficients			right-hand coefficient	Check column
m_{ij}	a_{i1}	a_{i2}	a_{i3}	a_{i4}	a_{i5}
1	4	2	3	6	15
$\frac{5}{4}$	5	−5	3	3	6
$\frac{3}{4}$	3	−1	−2	2	2
1		−7·5	−0·75	−4·5	−12·75
$\frac{1}{3}$		−2·5	−4·25	−2·5	−9·25
1			−4	−1	−5
Roots	1·025	0·575	0·25		
s_j	12	−4	4	11	23

Note that the numbers 15, 6 and 2 in the last (check) column are obtained by summing the coefficients in the corresponding rows of the table (excluding the m_{ij}). The number $-12\cdot75$ is obtained by evaluating $6-\frac{5}{4}(15)$ and is then checked against the corresponding row sum $(-7\cdot5-0\cdot75-4\cdot5)$. The number $-9\cdot25$ is obtained by evaluating $2-\frac{3}{4}(15)$ and is then checked against the row sum $(-2\cdot5-4\cdot25-2\cdot5)$. Similarly the number -5 is obtained by evaluating $-9\cdot25-\frac{1}{3}(-12\cdot75)$ and is then checked against the row sum $(-4-1)$. The numbers in the last (check) row are obtained according to their definition, that is, the 12 is obtained by summing 4, 5 and 3, the -4 is obtained by summing 2, -5 and -1, etc.

Then $\quad s_1x_1+s_2x_2+s_3x_3 = 12(1\cdot025)+(-4)(0\cdot575)+4(0\cdot25)$
$$= 11$$
$$= s_4.$$

Finally $\qquad s_1+s_2+s_3+s_4 = 12-4+4+11 = 23 = s_5.$

Note that since all the arithmetic required in the above example was carried out exactly (and so without introducing any round-off errors) all the checks have been satisfied exactly.

Example. Solve the following system of equations working to three decimal places throughout and incorporating the checking procedures.

$$2\cdot801x_1 - 1\cdot643x_2 + 2\cdot017x_3 = 3\cdot216$$
$$0\cdot762x_1 + 1\cdot372x_2 - 2\cdot631x_3 = 1\cdot835$$
$$1\cdot314x_1 - 4\cdot721x_2 + 3\cdot596x_3 = 1\cdot062.$$

Multipliers m_{ij}	left-hand coefficients			right-hand coefficient	Check column
	a_{i1}	a_{i2}	a_{i3}	a_{i4}	a_{i5}
1	2·801	−1·643	2·017	3·216	6·391
0·272	0·762	1·372	−2·631	1·835	1·338
0·469	1·314	−4·721	3·596	1·062	1·251
1		1·819	−3·180	0·960	~~0·400~~ −0·401
−2·172		−3·950	2·650	−0·446	−1·746
			−4·257	1·639	~~2·617~~ −2·618
Roots s_j	1·340	−0·145	−0·385		
	4·877	−4·992	2·982	6·113	8·980

$$s_1 x_1 + s_2 x_2 + s_3 x_3 = (4\cdot877)(1\cdot340) - 4\cdot992(-0\cdot145) + 2\cdot982(-0\cdot385)$$
$$= 6\cdot111 \text{ to three decimal places}$$
$$\simeq s_4.$$

$$s_1 + s_2 + s_3 + s_4 = 4\cdot877 - 4\cdot992 + 2\cdot982 + 6\cdot113 = 8\cdot980 = s_5.$$

EXERCISES

Apply the checking procedures to the solutions of the exercises given at the end of the last sub-section.

8.1.2. Multiple right-hand sides

If it is required to solve a system of equations $Ax = b$ for a given matrix A and a series of different right-hand-side vectors b, then a considerable amount of work can be avoided by including all the columns of right-hand sides in the one table and proceeding as before. If the system is $n \times n$ we first obtain a value of x_n corresponding to each right-hand-side vector and then back substitution has to be performed for each right-hand side to give all the values of $x_{n-1}, x_{n-2}, \ldots, x_2, x_1$.

Now if there are m right-hand-side vectors

$$\begin{pmatrix} a_{1n+1} \\ a_{2n+1} \\ \vdots \\ a_{nn+1} \end{pmatrix}, \begin{pmatrix} a_{1n+2} \\ a_{2n+2} \\ \vdots \\ a_{nn+2} \end{pmatrix}, \ldots, \begin{pmatrix} a_{1n+m} \\ a_{2n+m} \\ \vdots \\ a_{nn+m} \end{pmatrix}$$

to be considered, the ith element in the check column is $\displaystyle\sum_{j=1}^{n+m} a_{ij}$, that is, the sum of all the elements in the ith row. In the check row, the scalar product of the vector $(s_1, s_2, \ldots, s_n)$ with the solution vector corresponding to the kth right-hand-side vector must be equal to the sum of the elements of the kth right-hand-side vector, that is, to s_{n+k}, for each k.

The table for the general 3×3 system with three different right-hand-side vectors is shown below.

Multipliers m_{ij}	left-hand coefficients			right-hand coefficients			Check column a_{i7}
	a_{i1}	a_{i2}	a_{i3}	a_{i4}	a_{i5}	a_{i6}	
1	a_{11}	a_{12}	a_{13}	a_{14}	a_{15}	a_{16}	$a_{17} = \sum\limits_{j=1}^{6} a_{1j}$
$m_{21} = \dfrac{a_{21}}{a_{11}}$	a_{21}	a_{22}	a_{23}	a_{24}	a_{25}	a_{26}	$a_{27} = \sum\limits_{j=1}^{6} a_{2j}$
$m_{31} = \dfrac{a_{31}}{a_{11}}$	a_{31}	a_{32}	a_{33}	a_{34}	a_{35}	a_{36}	$a_{37} = \sum\limits_{j=1}^{6} a_{3j}$
1		$a_{22}^{(1)}$	$a_{23}^{(1)}$	$a_{24}^{(1)}$	$a_{25}^{(1)}$	$a_{26}^{(1)}$	$a_{27}^{(1)}$
$m_{32} = \dfrac{a_{32}^{(1)}}{a_{22}^{(1)}}$		$a_{32}^{(1)}$	$a_{33}^{(1)}$	$a_{34}^{(1)}$	$a_{35}^{(1)}$	$a_{36}^{(1)}$	$a_{37}^{(1)}$
1			$a_{33}^{(2)}$	$a_{34}^{(2)}$	$a_{35}^{(2)}$	$a_{36}^{(2)}$	$a_{37}^{(2)}$
Roots	$x_1^{(1)}$	$x_2^{(1)}$	$x_3^{(1)} = \dfrac{a_{34}^{(2)}}{a_{33}^{(2)}}$				
	$x_1^{(2)}$	$x_2^{(2)}$	$x_3^{(2)} = \dfrac{a_{35}^{(2)}}{a_{33}^{(2)}}$				
	$x_1^{(3)}$	$x_2^{(3)}$	$x_3^{(3)} = \dfrac{a_{36}^{(2)}}{a_{33}^{(2)}}$				
s_j	s_1	s_2	s_3	s_4	s_5	s_6	s_7

To within round-off errors we should have

$$s_1 x_1^{(1)} + s_2 x_2^{(1)} + s_3 x_3^{(1)} = s_4$$
$$s_1 x_1^{(2)} + s_2 x_2^{(2)} + s_3 x_3^{(2)} = s_5$$
$$s_1 x_1^{(3)} + s_2 x_2^{(3)} + s_3 x_3^{(3)} = s_6.$$

Also $s_1 + s_2 + s_3 + s_4 + s_5 + s_6$ should be exactly equal to s_7.

Example. Solve the set of simultaneous equations $A\boldsymbol{x} = \boldsymbol{b}_k$ in which

$$A = \begin{pmatrix} 5 & 1 & -3 \\ 1 & 2 & 4 \\ 3 & -1 & 2 \end{pmatrix}$$

for the three right-hand-side vectors

$$\boldsymbol{b}_1 = \begin{pmatrix} 1 \\ 0 \\ 0 \end{pmatrix}, \qquad \boldsymbol{b}_2 = \begin{pmatrix} 0 \\ 1 \\ 0 \end{pmatrix}, \qquad \boldsymbol{b}_3 = \begin{pmatrix} 0 \\ 0 \\ 1 \end{pmatrix}.$$

(Retain three decimal places throughout the calculations.)

Multipliers	left-hand coefficients			right-hand coefficients			Check column
m_{ij}	a_{i1}	a_{i2}	a_{i3}	a_{i4}	a_{i5}	a_{i6}	a_{i7}
· 1	5	1	−3	1	0	0	4
0·2	1	2	4	0	1	0	8
0·6	3	−1	2	0	0	1	5
1		1·8	4·6	−0·2	1	0	7·2
−0·889		−1·6	3·8	−0·6	0	1	2·6
1			7·889	−0·778	0·889	1	~~9·001~~ 9·000
Roots from back substitution	0·112	0·142	−0·099				
	0·014	0·267	0·113				
	0·141	−0·325	0·127				
s_j	9	2	3	1	1	1	17

For the checks we obtain

$$9(0{\cdot}112) + 2(0{\cdot}142) + 3(-0{\cdot}099) = 0{\cdot}995 \simeq 1 = s_4,$$
$$9(0{\cdot}014) + 2(0{\cdot}267) + 3(0{\cdot}113) = 0{\cdot}999 \simeq 1 = s_5,$$
$$9(0{\cdot}141) + 2(-0{\cdot}325) + 3(0{\cdot}127) = 1{\cdot}000 = s_6$$

and $\quad s_1 + s_2 + s_3 + s_4 + s_5 + s_6 = 9 + 2 + 3 + 1 + 1 + 1 = 17 = s_7$.

8.2. Determination of the inverse A^{-1} of a matrix A

Consider the $n \times n$ system $A\boldsymbol{x} = \boldsymbol{b}_k$ for the series of right-hand-side vectors $\boldsymbol{b}_k$ ($k = 1, 2, \ldots, n$) which are the successive columns of the $n \times n$ unit matrix, that is, $b_1, b_2, \ldots, b_n$ are respectively the n-dimensional vectors

$$\begin{pmatrix} 1 \\ 0 \\ \vdots \\ 0 \end{pmatrix}, \begin{pmatrix} 0 \\ 1 \\ 0 \\ \vdots \\ 0 \end{pmatrix}, \ldots, \begin{pmatrix} 0 \\ 0 \\ \vdots \\ 0 \\ 1 \end{pmatrix}.$$

Let the respective solution vectors be denoted by x_k $(k = 1, 2, \ldots, n)$. Then

$$Ax_1 = b_1 = \begin{pmatrix} 1 \\ 0 \\ \vdots \\ 0 \end{pmatrix}, \quad Ax_2 = b_2 = \begin{pmatrix} 0 \\ 1 \\ 0 \\ \vdots \\ 0 \end{pmatrix}, \ldots, \quad Ax_n = b_n = \begin{pmatrix} 0 \\ 0 \\ \vdots \\ 0 \\ 1 \end{pmatrix}.$$

and so $x_1 = A^{-1}b_1, \quad x_2 = A^{-1}b_2, \ldots, x_n = A^{-1}b_n.$

Now form the $n \times n$ matrix X whose columns are the vectors $x_1, x_2, \ldots,$ x_n, that is

$$\begin{aligned} X &= [x_1, x_2, \ldots, x_n] \\ &= [A^{-1}b_1, A^{-1}b_2, \ldots, A^{-1}b_n] \\ &= A^{-1}[b_1, b_2, \ldots, b_n] \\ &= A^{-1} \begin{pmatrix} 1 & 0 & \vdots & \vdots & 0 \\ 0 & 1 & \vdots & \vdots & 0 \\ \vdots & \vdots & \vdots & \vdots & \vdots \\ 0 & 0 & \vdots & \vdots & 1 \end{pmatrix} = A^{-1}I \end{aligned}$$

where I is the $n \times n$ unit matrix.

Hence $X = A^{-1}$ and so the separate solution vectors of $Ax = b_k$ are the columns of the matrix A^{-1}.

Example. Keeping three decimal places throughout the calculations determine an approximation to the inverse of the matrix

$$\begin{pmatrix} 5 & 1 & -3 \\ 1 & 2 & 4 \\ 3 & -1 & 2 \end{pmatrix}.$$

Using Gaussian elimination, we must solve the system of equations $Ax = b_k$ $(k = 1, 2, 3)$ for the three right-hand-side vectors

$$b_1 = \begin{pmatrix} 1 \\ 0 \\ 0 \end{pmatrix}, \quad b_2 = \begin{pmatrix} 0 \\ 1 \\ 0 \end{pmatrix}, \quad b_3 = \begin{pmatrix} 0 \\ 0 \\ 1 \end{pmatrix}.$$

and
$$A = \begin{pmatrix} 5 & 1 & 3 \\ 1 & 2 & 4 \\ 3 & -1 & 2 \end{pmatrix}.$$

This is precisely the problem which was solved on page 145. From the results obtained there we see that an approximation to the inverse of the given matrix is the matrix

$$\begin{pmatrix} 0.112 & 0.014 & 0.141 \\ 0.142 & 0.267 & -0.325 \\ -0.099 & 0.113 & 0.127 \end{pmatrix}.$$

EXERCISES

1. Use Gaussian elimination to obtain approximations to the inverse of the matrices of the systems in exercises 1, 2, 3, 4 at the end of §8.1. Hence obtain approximate solutions of the given systems and compare these with the solutions already obtained.

8.3. Pivoting

If all our calculations could be carried out exactly, that is, without introducing any round-off errors, then there would be no more to say about Gaussian elimination. However, in general, it will be neither possible nor practical to use exact arithmetic in our computations, and so great care must be taken to ensure that the accumulation of round-off errors does not become too great.

To illustrate the problems involved we shall consider the solution using Gaussian elimination of the rather trivial two-by-two system.

$$10^{-5}x_1 + x_2 = 0.6$$
$$x_1 + x_2 = 1.$$

Of course, this system could be solved easily without Gaussian elimination but we shall use Gaussian elimination as the difficulties which then arise are the same as those which arise when solving higher-order systems. We shall use four significant figures in the calculations.

$$m_{21} = \frac{1}{10^{-5}} = 10^5 = 1.000 \times 10^5$$

and so we obtain

$$(1.000 - 1.000 \times 10^5)x_2 = 1.000 - 6.000 \times 10^4.$$

Then, using four significant figures, we have

$$-1{\cdot}000 \times 10^5 x_2 = -6{\cdot}000 \times 10^4$$

and so
$$x_2 = 6{\cdot}000 \times 10^{-1}.$$

The back substitution then gives

$$x_1 = 10^5(0{\cdot}6 - x_2) = 0.$$

Now the true solution of the given system correct to five decimal places can easily be obtained and is $x_1 = 0{\cdot}400\ 00$, $x_2 = 0{\cdot}600\ 00$. Thus the result we have obtained above for x_1 is far from satisfactory.

Now let us solve the same system again, still using Gaussian elimination, but changing the order of the two given equations. Thus we start with the system

$$x_1 + x_2 = 1$$
$$10^{-5}x_1 + x_2 = 0{\cdot}6.$$

This time $m_{21} = 10^{-5}$ and so we obtain

$$(1{\cdot}000 - 1{\cdot}000 \times 10^{-5})x_2 = 0{\cdot}6 - 1{\cdot}000 \times 10^{-5}.$$

Hence, to four significant figures,

$$1{\cdot}000 x_2 = 0{\cdot}6000$$

and so
$$x_2 = 0{\cdot}6000.$$

Now the back substitution gives

$$x_1 = (1 - x_2)$$
$$= (1 - 0{\cdot}6000)$$
$$= 0{\cdot}400.$$

This result is much more satisfactory! We must now attempt to find out what went wrong with our first Gaussian elimination calculations so that we can, if possible, avoid falling into the same trap again.

In the first solution, because of the relative smallness of the pivot, to obtain the coefficients of the modified second equation we had to add to each of the old coefficients of x_2 and the constant term, quantities so large numerically that, using only four significant figures, the old coefficients themselves were completely negligible; that is, the information contained in the second equation was completely lost. The back substitution in the first equation then resulted in x_1 being evaluated as the difference of two "very nearly equal" numbers with a resulting loss of significant figures. In fact in this example x_1 was calculated as the difference of two numbers which were equal to four significant figures, and so no significant figures were obtained in the result.

In the second solution the pivot was large, and so the coefficients of the modified second equation were obtained by adding to the respective old coefficients quantities which were relatively small. Thus the information in the (then) second equation was not lost. The (then) first equation was, of course, used for the back substitution, so that the information contained in both of the given equations was utilized.

8.3.1. Partial pivoting

The most commonly used method for attempting to avoid the above difficulty is known as *partial pivoting*. It consists of choosing as pivot at each stage of the Gaussian elimination process the numerically largest coefficient in the left-hand column. This is equivalent to re-ordering, if necessary, the rows in each derived system; that is, to writing down the reduced equations in a different order.

As an illustration of the effect of partial pivoting, consider the solution of the following system.

$$0{\cdot}7x_1 + 46x_2 - 16x_3 = 13{\cdot}3$$
$$19x_1 - 13x_2 + 23x_3 = 14$$
$$31x_1 + 26x_2 + 11x_3 = 17.$$

(The exact solution of this system is $x_1 = -1$, $x_2 = 1$, $x_3 = 2$.)

Without using partial pivoting, and retaining three significant figures throughout the computations, we obtain the following:

m_{ij}	a_{i1}	a_{i2}	a_{i3}	a_{i4}	a_{i5}
1	0·7	46	−16	13·3	44
27·1	19	−13	23	14	43
44·3	31	26	11	17	85
1		−126 × 10	457	−346	−115 × 10
1·60		−201 × 10	720	−572	−186 × 10
1			−11·2	−18·4	−20·0 −29·6
Roots	−0·620	0·869	1·64		
s_j	50·7	59	18	44·3	172

The relatively large discrepancy in the element in the check column at the last stage of the elimination process indicates a dangerously large accumulation of round-off errors. Note too that the coefficients in the last row of the elimination process were obtained as differences of pairs of considerably larger numbers.

Now $\qquad$ $s_1 x_1 + s_2 x_2 + s_3 x_3 = 49.4$

which differs appreciably from the value 44·3 obtained for s_4. This is a further indication that the solution we have obtained is rather inaccurate.

Repeating this example using partial pivoting we obtain the results shown below. The pivot element at each stage of the process is printed in bold type instead of changing the order of the equations.

m_{ij}	a_{i1}	a_{i2}	a_{i3}	a_{i4}	a_{i5}
0·0226	0·7	46	−16	13·3	44
0·613	19	−13	23	14	43
1	**31**	26	11	17	85
1		**45·4**	−16·2	12·9	42·1
−0·637		−28·9	16·3	3·58	−9·10 −9·02
1			**5·98**	11·8	17·8
Roots	−0·978	0·987	1·97		
s_j	50·7	59	18	44·3	172

Here $\qquad$ $s_1 x_1 + s_2 x_2 + s_3 x_3 = 44.1 \simeq 44.3 = s_4.$

In this case, partial pivoting has resulted in a more accurate solution being obtained. It must, however, be pointed out that while partial pivoting will generally improve the accuracy of a solution (when working to a restricted number of significant figures) it cannot be guaranteed to do so in every case.

EXERCISE

Construct a flow diagram for Gaussian elimination with partial pivoting.

8.3.2. Complete pivoting

The success of partial pivoting suggests that perhaps we can do even better if we choose as pivot not merely the numerically greatest coefficient in the first column, but the numerically greatest coefficient in the whole matrix of coefficients at each stage. This is equivalent to reordering, if necessary, the rows and the columns in each derived system. The method using this technique is called *complete pivoting*. It is more awkward to record when the computations are being carried out manually, is more difficult to program for a computer, and is more time-consuming than partial pivoting on a computer, since a two-dimensional scan of coefficients has to be made at each stage to determine the numerically largest instead of a one-dimensional scan. Further, it does not necessarily lead to more accurate results, and may indeed give poorer ones in many cases. This latter situation arises because the product of the (exact unrounded) pivots is equal to a fixed number (namely the determinant of the matrix of the given system). Hence the use of too large pivots to begin with may result in the final pivot being extremely small.

Example. Solve the following system of equations using (i) partial pivoting and (ii) complete pivoting.

$$x_1 - 19x_2 + 6x_3 + 72x_4 = 21$$
$$16x_1 + 56x_2 + 21x_3 - 7x_4 = 47$$
$$2x_1 + 7x_2 + 4x_3 + 3x_4 = 7$$
$$x_1 + 5x_2 + 3x_3 + x_4 = 4.$$

(The exact solution of these equations is $x_1 = -1$, $x_2 = 2$, $x_3 = -2$, $x_4 = 1$.) (In both cases we shall retain three significant figures throughout the calculations although, because of round-off errors and cancellations, this will not always be three-figure accuracy.)

The pivot elements are in bold type in both cases.

(i)

m_{ij}	a_{i1}	a_{i2}	a_{i3}	a_{i4}	a_{i5}	a_{i6}
0·0625	1	−19	6	72	21	81
1	**16**	56	21	−7	47	133
0·125	2	7	4	3	7	23
0·0625	1	5	3	1	4	14
1		**−22·5**	4·69	72·4	18·1	72·7
0		0	1·38	3·88	1·12	6·38
−0·0667		1·5	1·69	1·44	1·06	5·69
0·690			1·38	3·88	1·12	6·38
1			**2·00**	6·27	2·27	10·5
1				**−0·446**	−0·446	−0·865 −0·892
Roots	−1·00	2·00	−2·00	1·00		
s_j	20	49	34	69	79	251

$$s_1x_1 + s_2x_2 + s_3x_3 + s_4x_4 \simeq 80$$

F

which does not differ too much from the value 79 obtained for s_5. The product of the pivots $\simeq 321$.

(ii)

m_{ij}	a_{i1}	a_{i2}	a_{i3}	a_{i4}	a_{i5}	a_{i6}	
1	1	−19	6	**72**	21	81	
−0·0972	16	56	21	−7	47	133	
0·0417	2	7	4	3	7	23	
0·0139	1	5	3	1	4	14	
1	16·1	**54·2**	21·6		49·0	141	
0·144	1·96	7·79	3·75		6·12	19·6	
0·0970	0·986	5·26	2·92		3·71	12·9	
0·776	−0·358		0·640		−0·936	~~−0·704~~	−0·654
1	−0·576		**0·825**		−1·04	~~−0·777~~	−0·791
1	**0·0890**				−0·129	~~−0·0402~~	−0·0400
Roots	−1·45	2·25	−2·28	1·10			
s_j	20	49	34	69	79	251	

$$s_1 x_1 + s_2 x_2 + s_3 x_3 + s_4 x_4 \simeq 80$$

which once again does not differ too much from the value of 79 obtained for s_5. Note, however, the dangerous accumulation of round-off error which is suggested by the later values in the a_{i6} column.

The product of the pivots $\simeq 287$.

On comparison with the exact solution of the given system it is clear that in this case the approximate solution using partial pivoting is better (for each of the values x_1, x_2, x_3 and x_4) than the approximate solution obtained using complete pivoting. As stated above, this is essentially because using complete pivoting resulted in the first two pivots being so large that the later ones were excessively small.

Note that it would appear that the products of the pivots in (i) and (ii) are not nearly equal. The discrepancy between the two values is, however, also due to the build-up of round-off errors, mainly in the solution using complete pivoting. (The determinant of the matrix of the given system is 312.)

EXERCISES

1. The system

$$0\cdot000\,100 x_1 + 1\cdot00 x_2 = 1\cdot00$$
$$1\cdot00 x_1 + 1\cdot00 x_2 = 2\cdot00$$

has the solution

$$x_1 = 1{\cdot}000\ 10, \quad x_2 = 0{\cdot}999\ 90$$

rounded to the number of decimal places shown. Solve this system by Gaussian elimination, rounding to two significant figures, using (i) no pivoting, (ii) partial pivoting. Compare the results with the more accurate solution given above.

2. Use Gaussian elimination with partial pivoting to obtain approximate solutions for the systems of equations given in exercises 2, 3, 4 at the end of § 8.1.

Compare your results with the solutions already obtained. Compute also the products of the pivot elements in each case and compare these with the corresponding products of the pivots when solutions were obtained without partial pivoting.

8.4. Ill-conditioning

An ill-conditioned system is one for which relatively small changes in the coefficients produce relatively large changes in the solution. It is one for which the solution is very sensitive to small changes in the coefficients, and so is one for which any uncertainties in the coefficients are considerably amplified in the solution. Such uncertainties will be present if, for example, the coefficients are the results of any observational measurements. Even if the coefficients are exact, small errors will be introduced during the calculations. Indeed it may happen that if the coefficients have been rounded or obtained experimentally (to a limited accuracy) then no meaningful solution can be obtained. The *exact* solution of the system with the rounded coefficients may have *no* significant figures in common with the exact solution of the same system with more accurate coefficients. This difficulty is not in the method of solution but is inherent in the system of equations itself. For such a system it is necessary to use many more significant figures in the calculations than are required in the final results.

The whole problem of ill-conditioning and its recognition is a very interesting but rather complex one. However, pivots which decrease in magnitude at each stage, gradually tending to zero, give an indication of ill-conditioning. It is worth mentioning too that while a system with a nearly singular matrix will be ill-conditioned, it does *not* follow that an ill-conditioned system will have a nearly singular matrix.

As an example of an ill-conditioned system consider the equations

$$x_1 + x_2 = 2$$

$$x_1 + 1{\cdot}0001x_2 = 2{\cdot}0001.$$

Clearly the exact solution of this system is $x_1 = 1$, $x_2 = 1$.

However, if we now look at the system

$$x_1 + x_2 = 2$$
$$x_1 + 1{\cdot}0001x_2 = 1{\cdot}9999$$

the exact solution is $x_1 = 3$, $x_2 = -1$.

Thus changing one of the coefficients of the original system by approximately 1 in 10 000 completely changes the solution.

Also, the system

$$x_1 + x_2 = 2$$
$$x_1 + 0{\cdot}9999x_2 = 2{\cdot}0001$$

has the exact solution $x_1 = 3$, $x_2 = -1$ and so changing another coefficient of the original system by approximately 1 in 5000 also changes the solution completely.

EXERCISE

Compare the *exact* solutions of the two following systems

$$x_1 + 0{\cdot}99x_2 = 1{\cdot}99$$
$$0{\cdot}99x_1 + 0{\cdot}98x_2 = 1{\cdot}97$$

and

$$x_1 + 0{\cdot}99x_2 = 2{\cdot}00$$
$$0{\cdot}99x_1 + 0{\cdot}98x_2 = 1{\cdot}97.$$

First-order ordinary differential equations

9.1. Introduction

$$y' = 2xy$$

is a simple example of a first-order ordinary differential equation. It can be solved analytically as follows. We have

$$\int \frac{dy}{y} = 2 \int x\,dx$$

and hence

$$\ln y = 2(\tfrac{1}{2}x^2) + k$$

where k is an arbitrary constant.

Therefore $\qquad \ln Ky = x^2 \quad$ where $\quad k = -\ln K.$

Hence $\qquad\qquad y = Ce^{x^2} \quad$ where $\quad C = \dfrac{1}{K}.$

Thus the general solution of the given differential equation is $y = Ce^{x^2}$ for arbitrary constants C. If now we are given the initial condition that $y = 2$ when $x = 0$, we obtain $C = 2$ and the corresponding particular solution of the given differential equation is $y = 2e^{x^2}$.

When we solve a differential equation numerically, we cannot obtain a general solution as above but only a particular solution subject to some initial condition. Further, this numerical solution will not take the form of a functional expression as above but will simply be a sequence of numerical values of the solution function corresponding to a sequence of values of the independent variable. For example, the numerical solution of the differential equation $y' = 2xy$ with initial condition $y = 2$ when $x = 0$ is a sequence of numerical values of $2e^{x^2}$ corresponding to a sequence of values of x.

The general problem we are concerned with in this chapter is that of obtaining numerical solutions of first-order ordinary differential equations of the form $y' = f(x, y)$ subject to the initial condition $y(x_0) = y_0$. Here x_0 and y_0 are given constants, and f is a given function such that the differential equation has a unique solution. Thus we are concerned with determining a sequence of numerical values of y, corresponding to a sequence of values of x. In what follows we shall assume that the values of x in the sequence are equally spaced. We shall thus obtain values of y at equally spaced values of x.

Now, in the special case when f is independent of y, we have

$$y' = f(x)$$

and so

$$\int_{x_0}^{x_1} y'dx = \int_{x_0}^{x_1} f(x)dx.$$

Hence

$$[y(x)]_{x_0}^{x_1} = \int_{x_0}^{x_1} f(x)dx.$$

Therefore

$$y(x_1) = y(x_0) + \int_{x_0}^{x_1} f(x)dx$$

$$= y_0 + \int_{x_0}^{x_1} f(x)dx.$$

Now $\int_{x_0}^{x_1} f(x)dx$ can be evaluated approximately using one of the methods of numerical integration described in Chapter 6. Hence we obtain an approximation y_1 to the function value $y(x_1)$. Similarly we can obtain approximations $y_2, y_3, \ldots,$ to the function values $y(x_2), y(x_3), \ldots$; that is, we can obtain an (approximate) numerical solution of the given differential equation and initial condition.

9.2. Euler's method

When f is not independent of y we have

$$y' = f(x, y)$$

and hence, as above,

$$y(x_1) = y_0 + \int_{x_0}^{x_1} f(x, y)dx.$$

The difficulty now is in the evaluation of $\int_{x_0}^{x_1} f(x, y)dx$. Since the value of y is only known at $x = x_0$, this is the only point at which the function $f(x, y)$ can be evaluated and so the integral cannot be approximated using the trapezoidal rule or Simpson's rule as in Chapter 6. However, if we assume that f is approximately constant in the interval $x_0 \leqslant x \leqslant x_1$, then $f(x,y) \simeq f(x_0, y_0)$ in this interval and so we have

$$y(x_1) \simeq y_0 + \int_{x_0}^{x_1} f(x_0, y_0)dx$$

$$= y_0 + f(x_0, y_0) \int_{x_0}^{x_1} dx$$

$$= y_0 + hf(x_0, y_0)$$

where $x_1 - x_0 = h$, that is, we have obtained the approximation y_1 to $y(x_1)$ where

$$y_1 = y_0 + hf(x_0, y_0).$$

Similarly then we obtain the numerical approximation y_2 to the value $y(x_2)$ of the solution y at x_2, where

$$y_2 = y_1 + hf(x_1, y_1) \quad \text{and} \quad x_2 = x_1 + h.$$

Next we can evaluate y_3 the numerical approximation to $y(x_3)$ and then y_4 and so on.

In general we have

$$y_{r+1} = y_r + hf(x_r, y_r) \quad \text{with} \quad x_{r+1} = x_r + h, r = 0, 1, 2, \ldots.$$

In this way we can generate a numerical approximation to the solution of the given differential equation at the points $x_r = x_0 + rh$ for $r = 1, 2, 3, \ldots$. This method is called *Euler's method*.

Graphically the method is illustrated in Figure 20. Starting at the point (x_0, y_0) the numerical solution moves along the tangent to the solution curve to the point (x_1, y_1). The error at this stage is $y(x_1) - y_1$.

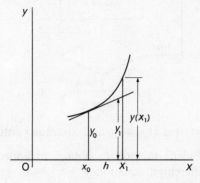

Figure 20

Now the slope of the tangent to the solution curve at the point (x_1, y_1) is calculated by evaluating $f(x_1, y_1)$ and the numerical solution proceeds by moving along this tangent to the point (x_2, y_2) where the process is again repeated and so on.

9.2.1. Flow diagram

A simple flow diagram for Euler's method is as follows:

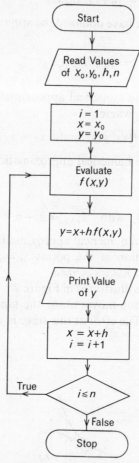

The solution is started at $x = x_0$ and continued until $x = x_0 + nh$.

Example. Use Euler's method to obtain an approximate numerical solution of the differential equation $y' = x^2 + 4x - \frac{1}{2}y$ with $y = 4$ when $x = 0$. (Use $h = 0.05$ and work to three significant figures throughout.)

Euler's method for the equation $y' = f(x, y)$ is given by

$$y_{r+1} = y_r + hf(x_r, y_r), \quad r = 0, 1, 2, \ldots.$$

Now for the given differential equation we have $f(x, y) = x^2 + 4x - \frac{1}{2}y$ and hence

$$y_{r+1} = y_r + 0.05(x_r^2 + 4x_r - \frac{1}{2}y_r).$$

Therefore
$$\begin{aligned}
y_1 &= y_0 + 0.05(x_0^2 + 4x_0 - \frac{1}{2}y_0) \\
&= 4 + 0.05(0^2 + 4 \times 0 - \frac{1}{2} \times 4) \\
&= 3.9.
\end{aligned}$$

Then
$$\begin{aligned}
y_2 &= y_1 + 0.05(x_1^2 + 4x_1 - \frac{1}{2}y_1) \\
&= 3.9 + 0.05((0.05)^2 + 4(0.05) - \frac{1}{2}(3.9)) \\
&\simeq 3.81.
\end{aligned}$$

Continuing in this way we obtain the results shown in the following table. Now this particular differential equation and initial condition have an analytic solution. Values of this analytic solution, correct to three significant figures, are given in the last column of the table so that they can be compared with the corresponding values obtained using Euler's method.

x_r	y_r	$f(x_r, y_r)$	$y(x_r)$
0	4	-2	4
0.05	3.9	-1.75	3.91
0.10	3.81	-1.50	3.82
0.15	3.73	-1.24	3.76
0.20	3.67	-1.00	3.70
0.25	3.62		3.65

EXERCISES

1. Use Euler's method to obtain an approximate numerical solution of the differential equation $y' = 2xy$ with $y = 2$ when $x = 0$.
   (Use $h = 0.05$ and work to three significant figures throughout.)

2. The method for obtaining an approximate numerical solution of the differential equation $y' = f(x, y)$ defined by

   $$y_{r+2} = y_r + 2hf(x_{r+1}, y_{r+1}) \quad r = 0, 1, 2, \ldots$$

   is known as the *mid-point rule*. Use this method to obtain an approximate numerical solution of the differential equation $y' = x^2 + 4x - \frac{1}{2}y$ with $y = 4$ when $x = 0$, and $y = 3.90$ when $x = 0.05$, and compare this solution with that obtained in the text using Euler's method.
   (Use $h = 0.05$; work to three significant figures throughout and continue the solution as far as $x = 0.25$.)

9.3. Truncation error in Euler's method

We must now attempt to obtain an estimate of some of the errors occurring during the application of Euler's method. The value obtained as an approximation to $y(x_1)$ is given by

$$y_1 = y_0 + hf(x_0, y_0).$$

However $y(x_1) = y(x_0 + h)$

$$= y(x_0) + hy'(x_0) + \frac{h^2}{2!} y''(x_0) + \ldots$$

$$= y(x_0) + hf(x_0, y(x_0)) + \frac{h^2}{2!} f'(x_0, y(x_0)) + \ldots.$$

Hence the error e_1 in our approximation is given by

$$e_1 = y(x_1) - y_1 = \frac{h^2}{2!} f'(x_0, y_0) + \cdots$$

if we assume that the initial value y_0 is equal to the exact function value $y(x_0)$. This error is called the *local truncation error* (at x_1) of the method and the first term $\frac{h^2}{2!} f'(x_0, y_0)$ in the series is called the *principal local truncation error* of the method. The error is local in the sense that it is the truncation error introduced due to the step from x_0 to x_1 (assuming that $y_0 = y(x_0)$).

Consider now the step from x_1 to x_2. We have

$$y_2 = y_1 + hf(x_1, y_1)$$

and $y(x_2) = y(x_1 + h) = y(x_1) + hy'(x_1) + \frac{h^2}{2!} y''(x_1) + \ldots$

$$= y(x_1) + hf(x_1, y(x_1)) + \frac{h^2}{2!} f'(x_1, y(x_1)) + \ldots.$$

Hence the error now is

$e_2 = y(x_2) - y_2$

$$= (y(x_1) - y_1) + h(f(x_1, y(x_1)) - f(x_1, y_1)) + \frac{h^2}{2!} f'(x_1, y(x_1)) + \ldots.$$

This error is the total truncation error of the method at x_2 and is in part due to the truncation error made at the stage x_1. If we assume that y_1 is equal to the exact function value $y(x_1)$, then we are left only with the truncation error introduced due to the step from x_1 to x_2, i.e. the local truncation error of the method at x_2. Hence the local truncation error at x_2 is

$$\frac{h^2}{2!} f'(x_1, y(x_1)) + \dots$$

$$= \frac{h^2}{2!} f'(x_1, y_1) + \dots$$

since $\quad f(x_1, y(x_1)) = f(x_1, y_1) \quad$ and $\quad f'(x_1, y(x_1)) = f'(x_1, y_1)$

when $y_1 = y(x_1)$.

The principal local truncation error at x_2 is

$$\frac{h^2}{2!} f'(x_1, y_1).$$

Similarly, the local truncation error of Euler's method at x_r $(r = 1, 2, \dots)$ is

$$\frac{h^2}{2!} f'(x_{r-1}, y_{r-1}) + \dots$$

and the principal local truncation error of the method at x_r is

$$\frac{h^2}{2!} f'(x_{r-1}, y_{r-1}).$$

Note that
$$f'(x, y) = \frac{\partial f}{\partial x} + \frac{\partial f}{\partial y} y'$$

$$= \frac{\partial f}{\partial x} + f \frac{\partial f}{\partial y}.$$

Example. Evaluate the principal local truncation errors at $x = 0{\cdot}05$ and $x = 0{\cdot}25$ in the solution of the equation

$$y' = x^2 + 4x - \tfrac{1}{2} y$$

with $y = 4$ at $x = 0$ given on page 159.

The principal local truncation error at x_r is

$$\frac{h^2}{2!} f'(x_{r-1}, y_{r-1}).$$

Here $\qquad\qquad h = 0{\cdot}05 \quad$ and $\quad f = x^2 + 4x - \tfrac{1}{2} y.$

Therefore
$$f' = 2x + 4 - \tfrac{1}{2}(x^2 + 4x - \tfrac{1}{2}y)$$
$$= -\tfrac{1}{2}x^2 + 4 + \tfrac{1}{4}y.$$

Hence the principal local truncation error at 0·05 is

$$\frac{(0\cdot05)^2}{2}(4 + \tfrac{1}{4}(4)) = 0\cdot00625.$$

The principal local truncation error at 0·25 is

$$\frac{(0\cdot05)^2}{2}(-\tfrac{1}{2}(0\cdot20)^2 + 4 + \tfrac{1}{4}(3\cdot67)) \simeq 0\cdot0061.$$

EXERCISES

1. Evaluate the principal local truncation errors at $x = 0\cdot05$ and $x = 0\cdot25$ in the solution of the equation $y' = 2xy$ with $y = 2$ when $x = 0$.
   (Use the solution obtained to exercise 1 at the end of the last section.)

2. Obtain an expression for the principal local truncation error of the mid-point rule given in exercise 2 at the end of the last section.

9.4. Effect of decreasing step size

It is clear that reducing the step size h will reduce the magnitude of the truncation error for any particular method. It does not follow however that a more accurate solution of the differential equation will necessarily be obtained after a few steps of the method. The reason for this is beyond the scope of this book.

Appendices

Appendices

Appendix 1

Taylor series and the binomial expansion

If a function f has continuous derivatives $f^{(r)}$ $(r = 1, 2, \ldots, n)$ on an interval I which contains a then, for all x in I,

$$f(x) = f(a) + \frac{x-a}{1!} f'(a) + \frac{(x-a)^2}{2!} f''(a) + \cdots + \frac{(x-a)^{n-1}}{(n-1)!} f^{(n-1)}(a) + R_n,$$

where the remainder term R_n is given by

$$R_n = \frac{1}{(n-1)!} \int_a^x f^{(n)}(t)(x-t)^{n-1} dt.$$

Proof. The method used is that of induction.

For $n = 1$,

$$f(a) + R_1 = f(a) + \int_a^x f'(t) dt$$
$$= f(a) + \{f(x) - f(a)\}$$
$$= f(x).$$

Hence the result stated is true for $n = 1$.

Now assume that the result is true when $n = k$, where $1 \leqslant k \leqslant n-1$; that is, assume that

$$f(x) = f(a) + \frac{x-a}{1!} f'(a) + \cdots + \frac{(x-a)^{k-1}}{(k-1)!} f^{(k-1)}(a) + R_k$$

$$= f(a) + \frac{x-a}{1!} f'(a) + \cdots + \frac{(x-a)^{k-1}}{(k-1)!} f^{(k-1)}(a)$$

$$+ \frac{1}{(k-1)!} \int_a^x f^{(k)}(t)(x-t)^{k-1} dt$$

Then $\quad f(x) = f(a) + \dfrac{x-a}{1!} f'(a) + \cdots + \dfrac{(x-a)^{k-1}}{(k-1)!} f^{(k-1)}(a)$

$$+ \frac{1}{(k-1)!} \left\{ \left[f^{(k)}(t) \left(\frac{-(x-t)^k}{k} \right) \right]_a^x + \frac{1}{k} \int_a^x f^{(k+1)}(t)(x-t)^k dt \right\}$$

$$= f(a) + \frac{x-a}{1!} f'(a) + \cdots + \frac{(x-a)^{k-1}}{(k-1)!} f^{(k-1)}(a)$$

$$+ \frac{(x-a)^k}{k!} f^{(k)}(a) + R_{k+1}$$

where $\qquad\qquad R_{k+1} = \dfrac{1}{k!} \displaystyle\int_a^x f^{(k+1)}(a)(x-t)^k dt.$

Thus the result is also true for $n = k+1$.

Hence, by induction, the result stated is true for all positive integers n.

Now, if f has continuous derivatives of arbitrarily high order and if R_n can be made as small as we please by taking n sufficiently large, then we write

$$f(x) = \sum_{n=0}^{\infty} \frac{(x-a)^n}{n!} f^{(n)}(a).$$

The infinite series on the right-hand side is *Taylor's series* expansion for the function f about a.

In particular, if we take $f(x) = (1+x)^p$ and $a = 0$ we obtain

$$(1+x)^p = 1 + \frac{p}{1!} x + \frac{p(p-1)}{2!} x^2 + \cdots + \frac{p(p-1)\cdots(p-n+1)}{n!} x^n + \cdots$$

which is the *binomial expansion* of $(1+x)^p$. This expansion is valid in general for $-1 < x < 1$. (If p is a non-negative integer, the expansion terminates after a finite number of terms and is valid for all real values of x.)

When $p = -1$, we obtain

$$(1+x)^{-1} = 1 - x + x^2 - x^3 + \cdots \quad \text{for} \quad -1 < x < 1.$$

Appendix 2

Here we prove that the nth differences of the nth degree polynomial

$$a_n x^n + a_{n-1} x^{n-1} + \cdots + a_1 x + a_0 \quad (a_n \neq 0)$$

have the constant value $n! a_n h^n$ in which h is the difference between successive values of x at which the polynomial is evaluated.

Consider the general first-degree polynomial $a_1 x + a_0$ $(a_1 \neq 0)$. Its difference table has the form

x	$f(x)$	
x_0	$a_1 x_0 + a_0$	
x_1	$a_1 x_1 + a_0$	$a_1(x_1 - x_0)$
x_2	$a_1 x_2 + a_0$	$a_1(x_2 - x_1)$
x_3	$a_1 x_3 + a_0$	$a_1(x_3 - x_2)$
x_4	$a_1 x_4 + a_0$	$a_1(x_4 - x_3)$

But $\qquad (x_1 - x_0) = (x_2 - x_1) = (x_3 - x_2) = (x_4 - x_3) = h$

and so the first differences have the constant value $a_1 h = 1! ah$. Thus the theorem is true for $n = 1$.

Now assume that the theorem is true for n equal to some positive integer m; that is, assume that the mth differences of an mth degree polynomial have the constant value $m! a_m h^m$. Consider the $(m+1)$th degree polynomial $p_{m+1}(x)$ given by

$$p_{m+1}(x) = a_{m+1} x^{m+1} + a_m x^m + \cdots + a_2 x^2 + a_1 x + a_0 \quad (a_{m+1} \neq 0).$$

Form the function $q(x) = p_{m+1}(x+h) - p_{m+1}(x)$.

Then $\quad q(x) = a_{m+1}\{(x+h)^{m+1} - x^{m+1}\} + a_m\{(x+h)^m - x^m\} + \cdots$
$$+ a_2\{(x+h)^2 - x^2\} + a_1\{(x+h) - x\}.$$

$$= (m+1)a_{m+1} h x^m + \text{terms of integral degree less than } m \text{ in } x.$$

Thus $q(x)$ is a polynomial of degree m and so by our assumption its mth differences will have the constant value $m!\{(m+1)a_{m+1}h\}h^m$, that is the value $(m+1)!a_{m+1}h^{m+1}$. But the values of the first differences of $p_{m+1}(x)$ are simply values of the polynomial $q(x)$. For example,

$$p_{m+1}(x_1) - p_{m+1}(x_0) = p_{m+1}(x_0+h) - p_{m+1}(x_0) = q(x_0).$$

Hence the $(m+1)$th differences of the polynomial $p_{m+1}(x)$ will be mth differences of the polynomial $q(x)$ and so will have the constant value $(m+1)!a_{m+1}h^{m+1}$. Thus, if the theorem is true for a positive integer m, it follows that it is also true for the positive integer $(m+1)$. But we have shown that the theorem is true for the positive integer 1 and so, by induction, it must be true for all positive integer values of n.

Appendix 3

Let
$$I = \int_{x_0}^{x_2} \left\{ f_0 + \frac{x-x_0}{h} \Delta f_0 + \frac{(x-x_0)(x-x_1)}{2h^2} \Delta^2 f_0 \right\} dx.$$

$$\int_{x_0}^{x_2} f_0 \, dx = f_0 [x]_{x_0}^{x_2} = 2hf_0.$$

$$\int_{x_0}^{x_2} \frac{x-x_0}{h} \Delta f_0 \, dx = \frac{1}{h} \Delta f_0 [\tfrac{1}{2}(x-x_0)^2]_{x_0}^{x_2}$$

$$= 2h\Delta f_0.$$

Using integration by parts we obtain

$$\int_{x_0}^{x_2} \frac{(x-x_0)(x-x_1)}{2h^2} \Delta^2 f_0 \, dx = \frac{1}{2h^2} \Delta^2 f_0 \left\{ [\tfrac{1}{2}(x-x_0)^2(x-x_1)]_{x_0}^{x_2} \right.$$

$$\left. - \int_{x_0}^{x_2} \tfrac{1}{2}(x-x_0)^2 \, dx \right\}$$

$$= \frac{1}{2h^2} \Delta^2 f_0 \{ 2h^3 - \tfrac{1}{6}[(x-x_0)^3]_{x_0}^{x_2} \}$$

$$= \frac{1}{2h^2} \Delta^2 f_0 (2h^3 - \tfrac{4}{3}h^3)$$

$$= \tfrac{1}{3} h \Delta^2 f_0.$$

Hence
$$I = 2hf_0 + 2h\Delta f_0 + \tfrac{1}{3}h\Delta^2 f_0$$
$$= \tfrac{1}{3}h\{6f_0 + 6(f_1 - f_0) + (f_2 - 2f_1 + f_0)\}$$
$$= \tfrac{1}{3}h(f_0 + 4f_1 + f_2).$$

Appendix 4

Consider the polynomial

$$p(x) = a_n x^n + a_{n-1} x^{n-1} + \cdots + a_1 x + a_0 \quad (a_n \neq 0).$$

We can obtain an estimate of ε, the change in a root λ due to a change α_r in one of the coefficients a_r, as follows. The quantity $(\lambda + \varepsilon)$ is a root of

$$a_n x^n + a_{n-1} x^{n-1} + \cdots + (a_r + \alpha_r) x^r + \cdots + a_1 x + a_0$$

that is of $p(x) + \alpha_r x^r$ so that

$$p(\lambda + \varepsilon) + \alpha_r (\lambda + \varepsilon)^r = 0.$$

But λ is a root of $p(x)$ so that $p(\lambda) = 0$.

Hence
$$p(\lambda + \varepsilon) - p(\lambda) + \alpha_r (\lambda + \varepsilon)^r = 0.$$

Therefore
$$(p(\lambda) + \varepsilon p'(\lambda) + \cdots) - p(\lambda) + \alpha_r (\lambda^r + r \lambda^{r-1} \varepsilon + \cdots) = 0.$$

Then, to the first order in the quantities ε and α_r, we have

$$\varepsilon p'(\lambda) + \alpha_r \lambda^r = 0$$

and so
$$\varepsilon = -\alpha_r \left(\frac{\lambda^r}{p'(\lambda)} \right).$$

Hence, if $\left(\dfrac{\lambda^r}{p'(\lambda)} \right)$ is large for a particular root λ then a relatively small change α_r in the coefficient a_r can produce quite a large change ε in the root λ, that is, the given polynomial is ill-conditioned for this root. (This indication of ill-conditioning is still useful even when, due to the ill-conditioning itself, ε is not small and so the first-order approximation made above is quantatively very inaccurate.) Clearly then ill-conditioning will occur with nearly equal roots because of the smallness of p' in the neighbourhood of the roots, but it can occur in other cases also. We also see from the above expression for ε that an equation can be ill-conditioned for some of its roots and not for others, and that the smaller roots will tend to be better conditioned than the larger ones. Great care must be exercised when dealing with ill-conditioned roots. In such cases it will be necessary to work with many more significant figures than are required in the roots. It must be emphasized too that ill-conditioning is inherent in the equation itself and will have an effect whatever method is being used to détermine the roots.

Answers

Chapter 2

Section 2.4, page 7

1.   (i) 1·75; round-off error 0·0014;
    (ii) 0·0480; −0·00005;
    (iii) 349; 0·1862;
    (iv) 29·35; −0·00375;
    (v) 0·4863; −0·000049;
    (vi) 1·649; −0·0004997;
    (vii) 0·000498; 0·0000005;
    (viii) 5·0; 0·0032.

2.   (i) 1·751; 0·0004;
    (ii) 0·048; −0·00005;
    (iii) 349·186; 0·0002;
    (iv) 29·3462; 0·00005;
    (v) 0·4863; −0·000049;
    (vi) 1·6485; 0·0000003;
    (vii) 0·000; 0·0004985;
    (viii) 5·00; 0·0032.

Section 2.6.4, page 19

1.   (i) 13·37 correct to two decimal places. Third decimal place is 4 or 5.
    (ii) 24·40 correct to two decimal places. Alternatively $28·403 \pm 0·001$.
    (iii) 11·04 correct to two decimal places. To three decimal places it is either 11·038 or 11·037.
    (iv) 8·68 correct to two decimal places. Third decimal place is 3 or 4.
    (v) −2·3 correct to one decimal place. Second decimal place is 4 or 5. Alternatively $−6·345 \pm 0·001$.
    (vi) 11·02 correct to two decimal places. Third decimal place is 0 or 1.
    (vii) No significant figures obtainable. Sign of number indeterminate.

2. Better to add numbers and then round-off the total.

3.   (i) 147·5. To two decimal places the true value may be 147·49, 147·50 or 147·51. That is $147·50 \pm 0·01$.
    (ii) 0·094. The fourth decimal place is 2, 3 or 4. Alternatively $0·0943 \pm 0·0001$.
    (iii) 345·9 or $345·92 \pm 0·03$.
    (iv) 1·934 or $1·9340 \pm 0·0002$.
    (v) 0·00678. To six decimal places the true value may be 0·006779 or 0·006780.

4. From tables $\sqrt{2\cdot14} = 1\cdot4629$ to four decimal places.

Approximate absolute error is $\dfrac{1}{2}\left(\dfrac{0\cdot005}{1\cdot5}\right) \simeq 0\cdot002$.

Hence $\sqrt{a} = 1\cdot463 \pm 0\cdot002$.
From tables $\sqrt{8\cdot27} = 2\cdot8758$ to 4 decimal places.

Approximate absolute error is $\dfrac{1}{2}\left(\dfrac{0\cdot005}{2\cdot9}\right) \simeq 0\cdot001$.

Hence $\sqrt{b} = 2\cdot876 \pm 0\cdot001$.

5. From six-figure tables $\sin 0\cdot359 \simeq 0\cdot351338$.
Approximate maximum absolute error is $0\cdot0005 \cos 0\cdot359 \simeq 0\cdot0005$.
Hence $\sin a = 0\cdot3513 \pm 0\cdot0005$.
From six-figure tables $\sin 0\cdot745 \simeq 0\cdot677972$.
Approximate maximum absolute error is $0\cdot0005 \cos 0\cdot745 \simeq 0\cdot0004$.
Hence $\sin b = 0\cdot6780 \pm 0\cdot0004$.

Section 2.7, page 22

1. $0\cdot0228$.

2. $17\cdot9$, $0\cdot0559$ (last digit not reliable. Approximate absolute error less than or equal to $0\cdot0002$).

3. -100, $0\cdot0400$.
Value of polynomial at $x = 100$ is -4 (quite different from zero!).
Value at $x = 0\cdot0400$ is $0\cdot0016$.
$(x+100)(x-0\cdot0400) = x^2 + 99\cdot96x - 4$ which is a good approximation to the given polynomial.

4. $\dfrac{1-\cos x}{\sin x} = \tan \tfrac{1}{2}x$.

5. $\dfrac{\dfrac{1}{\sqrt{x}} - \dfrac{1}{2}}{x-4} = \dfrac{-1}{2(2\sqrt{x}+x)}$. $-0\cdot0637$ to 3 significant figures.

Chapter 3

Section 3.1.1, page 27

1. $-0\cdot71875$.

2. $-6\cdot032$.

Section 3.2, page 29

1. $5x^3 + 13x^2 + 25x + 56$, 105.

2. $3x^6 - 8x^5 + 20x^4 - 58x^3 + 176x^2 + 527x - 1580$, 4739.

3. $-1\cdot66108$.

4. $1\cdot33994$.

5. $p(x) = \left(x - \dfrac{b}{a}\right)q(x) + R = (ax - b)\left(\dfrac{q(x)}{a}\right) + R.$

$0.5x^3 - 1.25x^2 - 4.875x + 8.4375,\ 21.5625.$

Chapter 4

Section 4.1.4, page 33

$\delta f_{-1/2} = -0.05,\ \delta^2 f_3 = 0.0040,\ \delta^4 f_{-2} = 0.1998,\ \Delta f_2 = -0.0179,\ \Delta^2 f_3 = 0.0028,$
$\Delta^3 f_{-3} = -0.0501,\ \nabla f_3 = -0.0179,\ \nabla^3 f_{-1} = -0.2499,\ \nabla^4 f_4 = 0.0017.$

Section 4.4, page 37

$3x^2 - 5x + 0.25.$

This is a bad fit to the given function at intermediate values of x.

Chapter 5

Section 5.1.2, page 46

1. $1.9217,\ 1.9102,\ 1.9256.$

2. $5.357.$

3. $5.956.$

4. $0.175.$

Section 5.2, page 48

1. $5.358.$ Quadratic interpolation gives better result.

2. $5.954.$ Quadratic interpolation gives better result.
$5.6143.$

3. $0.164.$ Quadratic interpolation gives better result.

Section 5.3, page 51

1. $14.93.$

2. $1.348.$

3. $2x^2 - 3x + 0.75.$

4. $-3x^2 + 4x + 1.25.$

Chapter 6

Section 6.1.5, page 67

1. $0.14752,\ 0.15216,\ 0.15332,\ 0.15361.$

2. (i) $0.70833.\ |\text{Round-off error}| \leqq 0.000005.\ |\text{Truncation error}| \leqq 0.04.$
(ii) $0.69702,\ 0.000005,\ 0.01.$
(iii) $0.69412,\ 0.000005,\ 0.003.$

3. Take $h = 0.09$ and retain four decimal places.
0.2624 (True value rounded to five decimal places is 0.26236).

Section 6.2.4, page 73
1. 1.4939, 1.4014, 1.3974.

2.   (i) 0.69445, 0.000005, 0.008.
  (ii) 0.69326, 0.000005, 0.0005.
  (iii) 0.69315, 0.000005, 0.00003.

3. Take $h = 0.15$ and retain four decimal places. 0.262.

Section 6.3, page 74
1. (i) Truncation error is negative and has magnitude $\leqq \frac{1}{12}h^2(1.6-0)e^{1.6} \simeq 0.026$.
Magnitude of round-off error $\leqq nh\frac{1}{2}10^{-k} \simeq 0.8 \times 10^{-k} = 0.0008$ rounding
to three decimal places. This is much less than the possible magnitude of
the truncation error and so there is no need to use any more than three
decimal places in the working.

Then $\displaystyle\int_0^{1.6} e^x dx \simeq 3.966$.

But since maximum truncation error is estimated by -0.026 we see that
the value of the integral lies between 3.97 and 3.94.

 (ii) Maximum truncation error is estimated by -0.00007.
Maximum round-off error (using all the available figures) is estimated by
0.00008.

Then $\displaystyle\int_0^{1.6} e^x dx \simeq 3.95305$.

Using the above error estimates we see that the value of the integral lies
between 3.95313 and 3.95290 and so is 3.953 to three decimal places.

2. Using the trapezoidal rule and Simpson's rule and successively halving the
interval size h we obtain in both cases the value 0.361.
(Analytically the true value of the integral rounded to four decimal places is
0.3614.)

3. 1.91. True value rounded to three decimal places is 1.912.

Section 6.4.2, page 84
1. $\frac{1}{2}h$, $\frac{1}{2}h$. (c.f. trapezoidal rule).

2. $\frac{1}{3}h$, $\frac{4}{3}h$, $\frac{1}{3}h$, (c.f. Simpson's rule).

3. $\frac{4}{5}$, $\frac{16}{15}$, $\frac{2}{15}$.

4. $\dfrac{h}{\sqrt{3}}$, $-\dfrac{h}{\sqrt{3}}$.

5. 0.181.
|Truncation error| is much less than $\frac{19}{720} \times 32 \times 10^{-5} \simeq 0.000008$.
|Round-off error| $< \frac{0.2}{24}(9+19+5+1)\frac{1}{2}10^{-4} \simeq 0.00001$.

Chapter 7

Section 7.2, page 88
1. 0.424.

2. Process cycles round $0.514, 0.508, 0.514, \ldots$.

Section 7.3.1, page 98
1. $[0, \frac{1}{2}]$.

2. $F'(x) = \dfrac{x-1}{x^3}$.

x^3 increases in $(0, \frac{1}{2}]$, $|x-1|$ decreases and so $|F'|$ decreases.
Hence $\min |F'|$ in $(0, \frac{1}{2}]$ is $|F'(\frac{1}{2})| = 4 > 1$. Therefore $|F'(\lambda)| > 1$.
Hence method is not suitable.
$x_0 = 0.4$, $x_1 = 0.625$, $x_2 = -0.320$, $x_3 = 8.008$.

3. $F'(x) = \dfrac{-1}{3(\frac{1}{2}-x)^{2/3}}$.

$(\frac{1}{2}-x)$ decreases and so $|F'|$ increases as x increases from $\frac{1}{3}$ to $\frac{1}{2}$.

Hence $\min |F'|$ in $[\frac{1}{3}, \frac{1}{2})$ is $|F'(\frac{1}{3})| = \dfrac{\sqrt[3]{36}}{3} > 1$.

Therefore $|F'(\lambda)| > 1$ and so the method is not suitable.
$x_0 = 0.4$, $x_1 = 0.464$, $x_2 = 0.330$, $x_3 = 0.554$, $x_4 = -0.378$, $x_5 = 0.958$.

4. $(0, \pi)$.

5. $f(-4) = -20, f(-3) = 7, f(0) = 4, f(1) = -5, f(2) = -8, f(3) = 1$.
Hence roots in stated intervals.

$$F'(x) = \tfrac{3}{10}x^2 \text{ which is} \begin{cases} < 1 \text{ for } x \in [0, 1] \\ > \frac{27}{10} > 1 \text{ for } x \in [-4, -3] \\ > \frac{12}{10} > 1 \text{ for } x \in [2, 3]. \end{cases}$$

Hence given formula is suitable for the root in $[0, 1]$ but not for the other two
$x_{n+1} = (10x_n - 4)^{1/3}$
$0.4067, -3.35, 2.94$.

6. $f(0) = \frac{1}{2}, f(\frac{1}{2}\pi) = \frac{3}{2} - \frac{1}{2}\pi < 0$.
Hence there is a root in $(0. \frac{1}{2}\pi)$.
$x_{n+1} = \sin x_n + \frac{1}{2}$.
1.497.

Section 7.5, page 107
1. $0.407, 2.94$.

2. 0.51.

Section 7.6, page 111
1. 0.424.

2. 0.51.

3. 2·94.

4. 0·407.

Section 7.7.6, page 121
1. 0·424.

2. 0·51.

3. 1·497.

4. 1·456.

5. 0·567.

7. $x_{n+1} = x_n(2 - ax_n)$.

8. 1·857, 4·536.

9. 0·787.

Section 7.8, page 126
1. 0·407, $-3·35$, 2·94.

2. $-0·170$, 1·689, 3·481.

3. 1·040, $-1·341$.

Section 7.9, page 131
1. 2·396, 1·576.

2. 0·768, 0·732.

Chapter 8

Section 8.1, page 139
1. 1·000, 2·000, $-1·000$.

2. 1·183, 0·035, 0·484.

3. 179, -138, $-45·5$.

4. 5·49, 1·01, $-0·0306$.

Section 8.2, page 147

1. Inverse
$$\begin{pmatrix} 0·1062 & 0·2062 & -0·01250 \\ -0·06233 & -0·06233 & 0·1250 \\ -0·1813 & 0·1187 & -0·03750 \end{pmatrix}$$

Roots   0·9996,   2·001, $-1·000$.

Inverse
$$\begin{pmatrix} 0·1145 & 0·2844 & -0·06913 \\ -0·07341 & -0·08769 & 0·08205 \\ -0·02804 & -0·2382 & 0·1490 \end{pmatrix}$$

Roots   1·1787,   0·0379,   0·4899.

Inverse $\begin{pmatrix} -1·04 & 0·314 & -0·0869 \\ 0·817 & 0·0486 & -0·307 \\ 0·296 & -0·141 & 0·108 \end{pmatrix}$

Roots   177, -138, $-45·5$.

Inverse $\begin{pmatrix} 2·14 & 1·26 & 0·123 \\ 0·465 & 0·277 & -0·00695 \\ 0·135 & -0·0466 & 0·00122 \end{pmatrix}$

Roots   4·53,   0·993, $-0·0280$.

Section 8.3.2, page 152
1. (i) 0·00, 1·00.
    (ii) 1·00, 1·00.

2. 1·181, 0·034, 0·483 (True solution 1·273080, 0·034557, 0·483801)

    175, -139, -443 (175·624, $-138·661$, $-44·6452$)
    0·0186, 1·23, $-0·0701$ (0·01866, 1·231, $-0·06993$)

Section 8.4, page 154
Solution of first system $x_1 = 1$, $x_2 = 1$.
Solution of second system $x_1 = -97$, $x_2 = 100$.

Chapter 9

Section 9.2.1, page 159

1.

x	y
0	2
0·05	2
0·10	2·01
0·15	2·03
0·20	2·06
0·25	2·10

2.

x	y
0	4
0·05	3·90
0·10	3·82
0·15	3·75
0·20	3·70
0·25	3·65

Section 9.3, page 162
1. 0·005, 0·006.

2. $\frac{1}{3}h^3 y'''(x_r)$.

Index